Rapid Assessment Program

A Biodiversity Assessment of the Eastern Kanuku Mountains, Lower Kwitaro River, Guyana

Editors
Jensen R. Montambault and
Olivier Missa

Bulletin
of Biological
Assessment

26

Center for Applied Biodiversity
Science (CABS)

Conservation International

Environmental Protection
Agency (EPA), Guyana

Centre for Study of Biological
Diversity (Biodiversity Centre)

Tropenbos International

The *RAP Bulletin of Biological Assessment* is published by:
Conservation International
Center for Applied Biodiversity Science
Department of Conservation Biology
1919 M St. NW, Suite 600
Washington, DC 20036
USA

202-912-1000 telephone
202-912-0773 fax
www.conservation.org
www.biodiversityscience.org

Editors: Jensen R. Montambault and Olivier Missa
Design/production: Kim Meek
Photography: Jensen R. Montambault and James Sanderson (camera-trap photos)
Map: Mark Denil
Translations: *Spanish:* Jensen R. Montambault and Ivia Martínez; *Portuguese:* Jensen R. Montambault and Inês Castro

***RAP Bulletin of Biological Assessment* Series Editors:**
Terrestrial and AquaRAP: Leeanne E. Alonso and Jennifer McCullough
Marine RAP: Sheila A. McKenna

ISBN: 1-881173-67-4
© 2002 by Conservation International
All rights reserved.
Library of Congress Catalog Card Number: 2002094101

RAP Bulletin of Biological Assessment was formerly *RAP Working Papers*. Numbers 1–13 of this series were published under the previous title.

Suggested citation: Montambault, J.R. and O. Missa (eds.). 2002. A Biodiversity Assessment of the Eastern Kanuku Mountains, Lower Kwitaro River, Guyana. RAP Bulletin of Biological Assessment 26. Conservation International, Washington, DC.

This publication was made possible through support provided by the Global Bureau, U.S. Agency for International Development, under the terms of the Biodiversity Corridor Planning and Implementation Program agreement, Award No. LAG-A-00-99-00046-00. The opinions expressed herein are those of the author(s) and do not necessarily reflect the views of the U.S. Agency for International Development, Conservation International, or other sponsoring organizations. The RAP survey was generously supported by The Rufford Foundation, The Smart Family Foundation, and USAID.

Table of Contents

Preface

Our planet faces numerous serious environmental problems, but at Conservation International (CI) we believe that, because of its irreversibility, one surpasses all others in terms of its importance: the extinction of biological diversity. Conservation efforts still receive only a tiny fraction of the human and financial resources needed to preserve biodiversity. As a result, we must use our resources efficiently, applying them to those areas with the highest concentration of biological diversity that is at most immediate risk of disappearing.

The Rapid Assessment Program (RAP) was conceived in 1989 to fill in the gaps in our knowledge of the 3–4% of the world's land surface "hotspots" which hold a third or more of the earth's terrestrial biodiversity and are considered at risk. The program has since grown to include freshwater and marine ecosystems, as well as a strong training element for regional partner scientists. RAP is a quick method of collecting scientific data needed to set conservation priorities and preserve biodiversity in mega-diverse areas. RAP results have already been applied directly to forming national parks in Bolivia and Perú and to developing a protected area strategy for Guyana. In addition, RAP findings helped to halt illegal oil drilling in a national park in Guatemala. Independent teams of regional scientists, trained by RAP, are now conducting long-term biological monitoring and additional rapid inventories in Brazil and Indonesia.

This RAP expedition followed recommendations made by the original RAP team in 1993 after completing a survey of the Western Kanuku Mountains. The scientific expedition was preceded by a two-week training course held at the Tropenbos Ecological Station, West Pibiri Creek, near the Mabura Hill Township, in which 17 Guyanese and Surinamese students received practical instruction on rapid biodiversity assessment methods. A team of RAP scientists from Canada, Suriname, Venezuela, and the United States, accompanied by five of the Guyanese students, then explored the little known flora and fauna in the region along the Lower Kwitaro River, on the eastern edge of the Eastern Kanuku Mountains. The taxonomic groups studied were plants, birds, fish, and mammals. Additional participants intended to inventory amphibians and reptiles were prevented from traveling to the RAP site due to restrictions following the September 11th terrorist attacks on the United States. We hope that the relationships formed between students, expert scientists, and Conservation International during this RAP training and expedition will continue to foster collaboration on environmental conservation projects both in the country of Guyana and throughout the Guayana Shield.

This report is intended for all interested conservationists, scientists, institutions, individuals, and organizations. We trust that the information presented here will catalyze effective conservation action on behalf of our planet's biodiversity, the legacy of life that is critical to us all.

Neville Waldron, Director, Conservation International-Guyana
Leeanne E. Alonso, Director, Rapid Assessment Program
Jensen R. Montambault, Manager, Rapid Assessment Program, Editor

Organizational Profiles

Conservation International

Conservation International (CI) is an international, nonprofit organization based in Washington, DC. CI believes that the Earth's natural heritage must be maintained if future generations are to thrive spiritually, culturally, and economically. Our mission is to conserve the Earth's living heritage, our global biodiversity, and to demonstrate that human societies are able to live harmoniously with nature.

Conservation International
1919 M Street NW, Suite 600
Washington, DC 20036
USA
Tel.   800-406-2306
Fax.   202-912-0772
Web.   www.conservation.org
        www.biodiversityscience.org

Conservation International–Guyana

266 Forshaw Street
Queenstown, Georgetown
Guyana
Tel.   592-225-2978; 592-227-8171; 592-223-5497
Fax.   592-225-2976

Lethem Field Office
Conservation International–Guyana
Lethem, Region 9
Guyana

Environmental Protection Agency (EPA), Guyana

The EPA was established by the Government of Guyana under the Environmental Protection Act of 1996 to "*coordinate and maintain a program for the conservation of biological diversity and its sustainable use.*" The objectives of the EPA include: implementing steps and systems to manage the natural environment and effectively ensure conservation; protecting and encouraging sustainable use of natural resources; preventing or controlling environmental pollution; and coordinating environmental management activities for individuals, organizations, and agencies (or any program with environmental content). The EPA also ensures sustainability by promoting public participation in the process of integrating environmental concerns in development planning, and coordinates a national environmental education and public awareness program.

The EPA developed the National Biodiversity Action Plan (NBAP), approved by the Cabinet in 1999. Programs under the NBAP are: wildlife management, bio-prospecting and research, *ex-situ* and *in-situ* conservation, developing a protected areas system and establishing a Protected Areas Secretariat, as well as ecotourism and the sustainable use of natural resources. The EPA is also responsible for current programs in environmental management (developing regulations, standards, and guidelines for environmental monitoring and enforcement); collaborative activities (including integrated coastal management, disaster preparation, waste management, air and water pollution, pesticide use, and monitoring climatic changes); and information, education, awareness, and capacity building programs.

Environmental Protection Agency (EPA)
IAST Building
University of Guyana
Turkeyen, Greater Georgetown
Guyana
Tel.   592-222-2277
Email.   epa@sdnp.org.gy

Centre for Study of Biological Diversity (Biodiversity Centre)

Since June 1992, the Biodiversity Centre has served as a repository for the Guyana National Herbarium and the University of Guyana's Zoological Museum. The Biodiversity Center is a joint project between the Smithsonian Institution, the University of Guyana, and World Wildlife Fund and assists the country's biodiversity management programs.

Biodiversity Centre activities include: curating, using, and maintaining plant and animal collections; facilitating scientific research by University of Guyana staff and students, other institutions, and interested members of the public; supporting educational activities with students of the University of Guyana and other institutions which encourage natural science research; and collaborating with other agencies and organizations for the conservation and sustainable use of the country's natural resources. Recently, the Biodiversity Centre has provided small grants for education and research accessible to the public, sponsored a local course in parataxonomy, established a GIS and Remote Sensing Laboratory with assistance from the World Bank, and promoted public awareness campaigns on biodiversity.

Centre for Study of Biological Diversity
University of Guyana
Turkeyen, Greater Georgetown
Guyana
Tel.   592-222-4921
Fax.   592-222-4921
Email.   csbd@guyana.net.gy

Tropenbos International

The Tropenbos Foundation was established in July 1988 as an expansion of the International Tropenbos Programme, developed in 1986 by the Government of The Netherlands. The Foundation contributes to the conservation and wise use of tropical rainforests through research and by developing methodologies. It involves and strengthens the capacity of local research institutions in relation to tropical rainforests.

To this end, research sites were established in several tropical countries. The Tropenbos-Guyana Programme, now obsolete, began in Guyana on September 1, 1989 to obtain a better understanding of lowland tropical rainforests ecosystems, and to use this information to achieve a sustainable forest management system. Under this system, successful timber harvesting (and possibly that of other non-timber forest products) would not lead to biodiversity degradation or the loss of proper hydrological functions in the exploited system.

To achieve these aims, Tropenbos conducted fundamental, applied, and extension studies in Guyanese forests with the cooperation of local organizations such as NARI, University of Guyana, Guyana Natural Resources Agency, and Utrecht University (The Netherlands). The primary research station was located in the Mabura Hill logging concession of the Demerara Timbers Limited (DTL), Central Guyana. Unexploited forest processes were analyzed at the field station in the Seeballi Compartment, and "gap studies", investigating the reaction of seedlings to biogeochemical changes in the forest after logging, were conducted at the West Pibiri Creek field station. In addition, anthropology and non-timber forest product research was conducted in the North-West District, Region 1.

The Guyana program was terminated in December 2001, and this agenda has become the responsibility of a local Guyana Forestry Commission research unit. Simultaneously, the Tropenbos Foundation was transformed into Tropenbos International (TBI) with a new mission statement aimed at ensuring that research includes the human use of forest resources in cases where populations are dependent on the forest.

Tropenbos International
P.O. Box 232
6700 AE Wageningen
The Netherlands
Email.   tropenbos@tropenbos.agro.nl
Web.   www.tropenbos.nl

Acknowledgments

Successful Rapid Assessment Program (RAP) expeditions are a product of the dedication, collaboration, and cooperation of many individuals and organizations, and the 2001 Guyana RAP training and expedition was no exception. First, we would like to acknowledge the Government of Guyana and affiliated agencies, including the Ministry of Amerindian Affairs, Environmental Protection Agency (EPA), the Region 9 Regional Democratic Council, as well as the Centre for the Study of Biological Diversity, for providing support and permitting and facilitating our work in the country's beautiful interior. Additional thanks are extended to the Guyana Defense Force (GDF) for providing RAP coordinators with jungle survival and First-Aid training before the expedition.

The Tropenbos-Guyana Programme and Demerara Timbers Limited (DTL) opened their doors to allow us to conduct the RAP methods training sessions at the Tropenbos Ecological Research Station at West Pibiri Creek near the Mabura Hill Township. Special thanks to Beverly and Maureen Daniels and Wendy Allicock for excellent cooking at West Pibiri, and also to Sandra James of Sandra's Shop, Mabura Hill for her invaluable role of replenishing both our scientific and cooking supplies during training. For transportation to and from the training site, the RAP team would like to recognize Andy and Alan Butler for their patience, tolerance, and skills in navigating the roads from Georgetown to West Pibiri and, most significantly, to Annai.

The RAP team would like to express its appreciation for the warm hospitality extended to the group by the management and staff at Rock View Resort in Annai, as well as the Community Development Chairman and residents of Rewa. We are deeply indebted to Duane de Freitas, Sr. and his Dadanawa Ranch crew of field guides, boat captains, and bowmen for their support, guidance, and innovation during the expedition. In particular, Andy Narine and Eloise Melville provided a tasty table under rough conditions.

Many thanks to Ray and Colin Gittens for bringing us safely from Annai to Lethem, and to Wesley Hendricks and Paul Liverpool for equipment shipping services between Lethem and Georgetown. The kind staff of the Takatu Guesthouse provided space and energy for the RAP Team during the report writing session at Lethem. Finally, we gratefully acknowledge Trans-Guyana Airways for returning the group safely to Georgetown.

The RAP team and Conservation International (CI)-Guyana would like to thank the following institutions for active participation in the Guyana RAP training and expedition: the EPA, Iwokrama International Centre for Rain Forest Conservation and Development, CI-Suriname, STINASU (Suriname), Tropenbos-Guyana Programme, Guyana Marine Turtle Conservation Society, Dadanawa Ranch, Centre for the Study of Biological Diversity, Rupununi Weavers Society, Royal Ontario Museum (Canada), University of Suriname, and the Jardín Botánico del Orinoco (Venezuela). We regret that additional members of the RAP team were unable to travel and participate in this scientific adventure due to the tragic terrorist attacks on the United States during this time. CI looks forward to the continued cooperation and support of these organizations for subsequent RAP biological surveys in Guyana.

For scientific field work and identification both in the field and the laboratory, RAP team members wish to thank the following individuals and organizations: Duane de Freitas, Sr.,

Ashley Holland, Guy Rodrigues, Duane de Freitas, Jr., Justin de Freitas, Asaph Wilson, Jonah Simon, Julian James, Ignacio Rufino, and additional personnel from Dadanawa, Karanambo, and Shea Village in the Rupununi Savannahs for providing invaluable field assistance, transportation, and information on local wildlife. We wish to recognize Angelina Licata and Gerardo Aymard, of the University Herbarium (PORT) at the Universidad Nacional Experimental de los Llanos Ezequiel Zamora (UNELLEZ), in Guanare, Portuguesa, Venezuela and specialists at the University of Wisconsin Herbarium and the Marie Selby Botanical Gardens Herbarium in Sarasota, FL for helping to identify botanical specimens. Finally, the researchers wish express their appreciation for the logistical arrangements made by the staff at the CI-Guyana and Washington offices for smooth organization of the expedition to the Eastern Kanuku Mountains, especially Eustace Alexander, Polyana Desa, Chris Desa, Kelvin Harris, Bernard de Souza, Neville Waldron, Joseph Singh, Lisa Famolare, Regina de Souza, Jenny Chun, Leslie Rice, Coreen Kopper, Jensen Montambault, and Olivier Missa.

This report is published as part of Conservation International's series, *RAP Bulletin of Biological Assessment*, and the editorial team would like to thank Leeanne Alonso, Kim Meek, Peter Hoke, Jennifer McCullough, Kayce Casner, Anthony Rylands, Jennifer Pervola Fermín, Eustace Alexander, Lisa Famolare, Bernard de Souza, Mark Engstrom, Graham Watkins, David Clarke, Tom Hollowell, Carol Kelloff, Aaron Bruner, Matt Foster, Inês Castro, and Ivia Martinez for their assistance in the editorial process.

This Rapid Assessment Program expedition was made possible by the continued and generous support of the Smart Family Foundation and the Rufford Foundation. The overflight, reconnaissance trips, and publication of this report were financed by the Global Bureau of the United States Agency for International Development (USAID) as part of a large-scale ecological corridor program with Conservation International. The Royal Ontario Museum (ROM) provided additional funding and supplies for a thorough small mammal study. The CI-Guianas regional program is grateful for the overall support from the Moore Foundation, W. Alton Jones Foundation, Adam and Rachel Albright, West Wind Foundation, The Ledder Foundation, Healthy Communities Initiatives, UNDP, UNESCO, and the Global Conservation Fund for making biodiversity projects in this region possible. Finally, we would like to thank Russ Mittermeier, Peter Seligmann, and Gustavo Fonseca for their continued support of the Rapid Assessment Program.

Participants and Authors

Eustace Alexander (Domestic Coordination)
Conservation International–Guyana
266 Forshaw Street
Queenstown, Georgetown
Guyana

Duane de Freitas, Sr. (Field Coordination)
Dadanawa Ranch
South Rupununi
Region 9
Guyana

Justin de Freitas (Fishes)
Dadanawa Ranch
South Rupununi
Region 9
Guyana

Wilmer Díaz (Plants)
Jardín Botánico del Orinoco
Módulos Laguna El Porvenir, Calle Bolívar
Ciudad Bolívar, Edo. Bolívar
Venezuela

Davis W. Finch (Birds)
WINGS
1643 North Alvernon Way, Suite 105
Tucson, AZ 85712
USA

Wiltshire Hinds (Birds)
Centre for the Study of Biological Diversity
University of Guyana
Turkeyen, Greater Georgetown
Guyana

Leroy Ignacio (Non-Volant Mammals)
Shulinab Village
South Rupununi
Region 9
Guyana

Burton K. Lim (Bats and Small Mammals)
Centre for Biodiversity and Conservation Biology
Royal Ontario Museum
100 Queen's Park
Toronto, Ontario M5S 2C6
Canada

Olivier Missa (Team Leader, Editor)
Avenue de l'universite, 88
B-1050 Brussels
Belgium
olimissa@yahoo.com

Jan H. Mol (Fishes)
University of Suriname
Center of Agricultural Research in Suriname (CELOS)
Leysweg, Paramaribo
Suriname

Jensen R. Montambault (Intl. Coordination, Editor)
Conservation International
Center for Applied Biodiversity Science
1919 M Street NW, Suite 600
Washington, DC 20036
USA

Zacharias Norman (Bats and Small Mammals)
Wowetta
Upper Takatu-Upper Essequibo
Region 9
Guyana

Jim Sanderson (Non-Volant Mammals)
Conservation International
Center for Applied Biodiversity Science
1919 M Street NW, Suite 600
Washington, DC 20036
USA

Corletta Toney (Field Assistant)
194 Lethem
Region 9
Guyana

Report at a Glance

Expedition Dates

20–29 September 2001

Area Description

The Kanuku Mountain range is located in the Rupununi region of southwestern Guyana. The RAP survey area focused on the lowland seasonally inundated and *terra firma* evergreen tropical rainforest along the Lower Kwitaro and Rewa rivers at the eastern edge of the Eastern Kanuku Mountain range. These rivers are part of the larger Essequibo River watershed. In some years, flooding in the surrounding Rupununi savannahs mixes with Brazil's Branco River savannahs to combine biological elements of both the Guayana Shield and Amazon catchments. High biodiversity and low human habitation makes this area ideal for conservation initiatives.

Reason for the Expedition

The Kanuku Mountains are presently under no legal protection status, although scientific surveys have already documented: (1) a high regional vertebrate diversity, particularly for birds and mammals; (2) healthy populations of many species threatened in other parts of the world such as the harpy eagle (*Harpia harpyja*), giant river otter (*Pteronura brasiliensis*), giant arapaima (*Arapaima gigas*), black caiman (*Melanosuchus niger*), giant armadillo (*Priodontes maximus*), giant anteater (*Myrmecophaga tridactyla*), and giant river turtle (*Podocnemis expansa*); (3) diverse habitats, from swamp forest and savannah along the Rupununi River to cloud forest on the Kanuku mountaintops; and (4) high tree species richness in the plant communities that combine Guayana Shield and Amazon elements.

Information on the region's biodiversity is scarce, however, making it difficult for policy makers to delineate protected areas or sometimes even justify conservation action. In 1993, Conservation International led a team of RAP scientists to explore the Western Kanuku Mountains and the Rewa River, and their impressive results suggested that the Eastern Kanukus would be perhaps even richer in biodiversity.

Major Results

The results from this RAP expedition along the Lower Kwitaro River on the eastern edge of the Eastern Kanuku Mountains confirm that this area contains rich biodiversity and is critically important for conservation on a national and international level. Mammal species richness was high; the results of our bat survey bring the region's total up to an impressive 89 bat species. Overall, the Kanukus contain about 70% of all mammals and 53% of all birds known to exist in Guyana. The Rupununi River basin maintains one of the richest fish faunas on earth, and our preliminary results show that the Kwitaro and Rewa rivers follow this trend. Plant communities were confirmed to be among the most diverse in Guyana and show little sign of human disturbance.

Number of Species

Plants (preliminary): 215 species
Birds: 264 species
Fish: 113 species
Non-volant Mammals: 25 species
Bats: 27 species

New Records for the Eastern Kanuku Mountain Region

Plants (preliminary): 40 species
Birds: 63 species
Fish: 113 species
Non-volant Mammals: 1 species
Bats: 4 species

Conservation Recommendations

The results of this RAP expedition show that it is critical to
place the Kanuku Mountain region, including both savannah
and forest habitat, and their microhabitat variations, under
some status of protection. This would effectively protect a
high proportion of the plant, bird, fish, and mammal species
of Guyana. The Eastern Kanuku Mountains are more suited
for developing a national park or strict protected area since
its human population is much lower than in the Western
Kanukus. The latter would be better for establishing an integrated management reserve with different levels of human
use and sustainable exploitation.

Informe a un Vistazo

Fechas de la Expedición
20–29 de septiembre de 2001

Descripción del Área
La Cordillera Kanuku está ubicada en la zona Rupununi de la región sur-oeste de Guayana. Nuestra investigación se concentró en el terreno bajo, inundado por temporada y *terra firma*, de bosque húmedo tropical y siempre-verde a la orilla del Bajo Río Kwitaro y el Río Rewa en el extremo este de la Cordillera Kanuku Oriental. Estos ríos son parte de la cuenca mayor del Río Essequibo, y en ciertos años las sabanas Rupununi que los rodean se sumergen y mezclan con aguas diluviales de las sabanas del Río Branco de Brasil. Esta inundación causa una combinación de los elementos de ambas bacías Guayanés y Amazonía. Un alto nivel de biodiversidad y bajo número de habitantes resulta en un área ideal para desarrollar iniciativas de conservación.

Razones para la Expedición
La Cordillera Kanuku actualmente no está bajo ningún estado legal de protección, aunque estudios científicos ya han documentado lo siguiente: (1) una alta diversidad de vertebrados al nivel regional; (2) poblaciones sanas de varias especies que están amenazadas en otras partes del mundo como la [usando nombres populares venezolanos] águila arpía (*Harpia harpyja*), perro del agua (*Pternura brasililiensis*), armadillo gigante (*Priodontes maximus*), oso hormiguero mayor (*Myrmecophaga tridactyla*), pirarucú (*Arapaima gigas*), tortuga arrau (*Podocnemis expansa*), y caimán negro (*Melanosuchus niger*); (3) diversos hábitats, desde el bosque pantanoso y la sabana a la orilla del Río Rupununi hasta el bosque nubloso encima de las montañas de la Cordillera Kanuku; y (4) una alta riqueza de especies de árboles en las comunidades de vegetación cuales combinan elementos del Escudo Guayanés y la Amazonía.

No obstante, la información acerca de la biodiversidad es escasa, y esto contribuye a la dificultad que tienen los encargados de la política nacional para delinear áreas protegidas o hasta justificar acciones conservacionistas. En 1993, Conservation International guió un equipo de científicos RAP para explorar la Cordillera Kanuku   Occidental y el Río Rewa. Los resultados de esta expedición sugieren que la Kanuku Oriental tendría una biodiversidad aún más rica.

Resultados Mayores
Los resultados de esta expedición RAP a la orilla del Bajo Río Kwitaro en el extremo este de la Cordillera Kanuku Oriental confirmaron que esta zona mantiene una biodiversidad muy rica y es crítica para la conservación ambiental al nivel nacional e internacional. La riqueza de especies de mamíferos fue alta, y los resultados de nuestra investigación de los murciélagos aumenta el total para la región a un número impresionante de 89 especies de munciélagos. Por lo general, las Kanukus contienen aproximadamente 70% de todos los mamíferos y 53% de todas las aves registrados en Guayana. La Cuenca del Río Rupununi mantiene una de las faunas de peces mas variadas en el mundo, y nuestros resultados preliminares muestran que los Ríos Kwitaro y Rewa siguen esta tendencia. Se confirmó que las comunidades de vegetación están entre las más

diversas en Guayana y han sufrido muy poca perturbación humana.

Número de Especies

Plantas (preliminar):	215 especies
Aves:	264 especies
Peces:	113 especies
Mamíferos no volantes:	25 especies
Murciélagos:	27 especies

Nuevos Registros para la Región de la Cordillera Kanuku Oriental

Plantas (preliminar):	40 especies
Aves:	63 especies
Peces:	113 especies
Mamíferos no volantes:	1 especie
Murciélagos:	4 especies

Recomendaciones para la Conservación

Los resultados de esta expedición RAP indican que es crítico poner la región de la Cordillera Kanuku, inclusivo los hábitats del bosque, sabana, y las variaciones de micro-hábitats, bajo protección legal. Esta acción efectivamente protegería una proporción alta de las especies de mamíferos, aves, peces, y plantas en Guayana. La Cordillera Kanuku Oriental esta en mejor condición para desarrollar un parque nacional o área protegida muy estricta porque su población humana es menor que en la Kanuku Occidental, donde sería mejor establecer una reserva de manejo integrado con diferentes niveles del uso humano y la explotación sostenible.

Relatório em Relance

Datas da Expedição
20–29 de setembro de 2001

Descrição do Local
A Cordilheira Kanuku esta situada na região de Rupununi no sudoeste da Guiana. Nosso levantamento concentrou-se na planície, de mata úmida tropical, inundada sazonalmente e *terra firma*, ao longo do baixo Rio Kwitaro e o Rio Rewa pela extremidade leste da Cordilheira Kanuku Oriental. Estes rios formam uma parte da bacia maior do Rio Essequibo. Durante certos anos, as inundações combinam elementos biológicos de ambas bacias das Guianas e Amazonas. Alta biodiversidade e baixa população humana fazem com que essa área seja ideal para iniciativas da conservação.

Razão da Expedição
A Cordilheira Kanuku não esta sob nenhum status de proteção legal, embora levantamentos científicos mostrarem o seguinte: (1) uma alta diversidade de vertebrados, especialmente de mamíferos e aves; (2) populações saudáveis de muitas espécies que estão ameaçadas em outras partes do mundo, como a gavião de penacho (*Harpia harpyja*), ariranha (*Pteronura brasiliensis*), preguiça de coleira (*Priodontes maximus*), tamanduá-bandeira (*Myrmecophaga tridactyla*), pirarucu (*Arapaima gigas*), jacaré-açu (*Melanosuchus niger*), e tartaruga da Amazônia (*Podocnemis expansa*); (3) hábitats diversos entre a mata pantanosa e savanas ao longe do Rio Rupununi até a mata de ncblina encima das montanhas Kanuku; e (4) alta riqueza de espécies de árvores entre as comunidades que combinam elementos do Escudo da Guiana e a Amazônia.

Não obstante, informação sobre essa biodiversidade é escassa, e por isso é difícil que os políticos possam delinear áreas protegidas o, às vezes, justificar as ações da conservação. Em 1993, Conservation International conduziu uma equipe de cientistas de RAP para explorar a Cordilheira Kanuku Ocidental e o Rio Rewa, e seus resultados impressionantes sugeriam que as Kanukus Orientais são as mais ricas da biodiversidade.

Resultados Maiores
Os resultados dessa expedição de RAP ao longe do baixo Rio Kwitaro pelo extremo leste da Cordilheira Kanuku Oriental confirmam que essa área mantém uma biodiversidade rica e que é crítica para conservação ao nível nacional e internacional. Riqueza de espécies de mamíferos foi alta, e os resultados de nosso levantamento de morcegos aumentam o total de espécies conhecidas na região para 89 que é muito impressionante. No geral, as Kanukus contém aproximadamente 70% de todos mamíferos e 53% de todas aves ocorrentes na Guiana. A Bacia do Rio Rupununi mantém uma das comunidades de peixes mais diversas no mundo, e nossos resultados preliminares confirmam essa tendência. Confirmou-se também que as comunidades de plantas têm a maior diversidade na Guiana e mostram muito pouca perturbação humana.

Espécies

Plantas (preliminar):	215 espécies
Aves:	264 espécies
Peixes:	113 espécies
Mamíferos não volantes:	25 espécies
Morcegos:	27 espécies

Novos Registros para a Cordilheira Kanuku Orient

Plantas (preliminar):	40 espécies
Aves:	63 espécies
Peixes:	113 espécies
Mamíferos não volantes:	1 espécies
Morcegos:	4 espécies

Recomendações pela Conservação

Os resultados desta expedição de RAP mostram que é necessário situar Cordilheira Kanuku, com ambos hábitats de savana e mata e suas variações de micro-hábitat, sob algum status de proteção. Efetivamente, isso protegeria a uma alta proporção das espécies de mamíferos, aves, peixe, e plantas da Guiana. As Kanukus Orienteis são mais apropriadas para implementação de parque nacional ou área estritamente protegida, uma vez que sua população humana é mais baixa que nas Kanukus Ocidentais que, por sua vez, são melhores para estabelecer uma reserva de manejo integrado com níveis diferentes de uso humano e utilização sustentável.

Eastern Kanuku Mountains

Lower Kwitaro River Guyana

RAP Expedition

17 September through 1 October, 2001

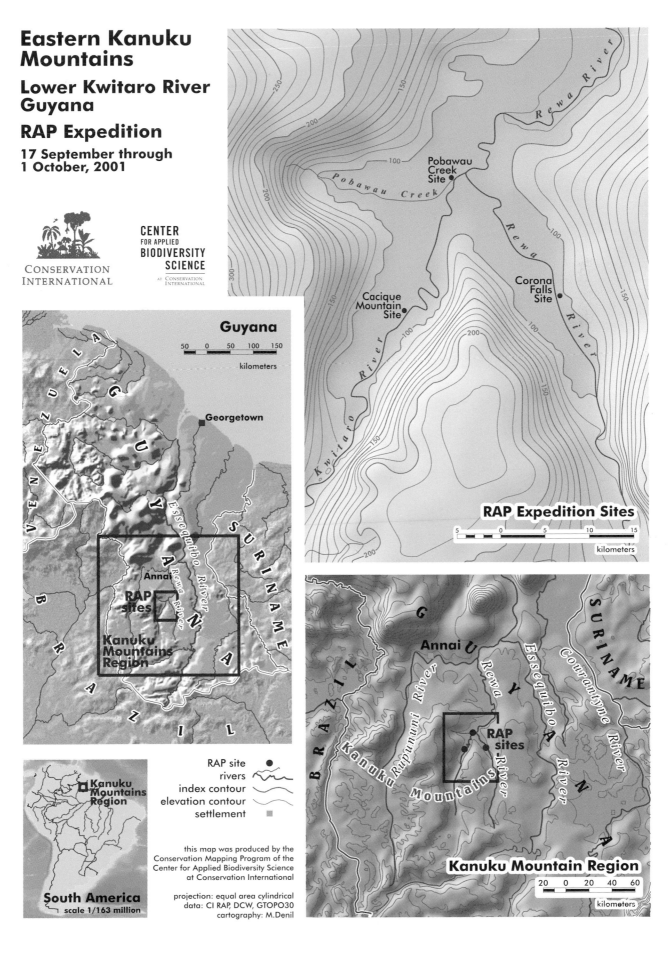

CONSERVATION INTERNATIONAL

CENTER FOR APPLIED BIODIVERSITY SCIENCE
AT CONSERVATION INTERNATIONAL

RAP Expedition Sites

Pobawau Creek Site

Corona Falls Site

Cacique Mountain Site

Rewa River

Pobawau Creek

Kwitaro River

kilometers

Guyana

Georgetown

VENEZUELA

GUYANA

SURINAME

BRAZIL

Annai

RAP sites

Kanuku Mountains Region

Essequibo River

Rewa River

kilometers

RAP site ●
rivers
index contour
elevation contour
settlement ■

Kanuku Mountains Region

South America
scale 1/163 million

this map was produced by the Conservation Mapping Program of the Center for Applied Biodiversity Science at Conservation International

projection: equal area cylindrical
data: CI RAP, DCW, GTOPO30
cartography: M.Denil

Kanuku Mountain Region

Annai

RAP sites

BRAZIL

GUYANA

SURINAME

Kanuku Mountains

Rupununi River

Rewa River

Essequibo River

Courantyne River

kilometers

Rear (L to R): Damian Fernandes, Hemchandranauth Sambhu, Justin de Freitas, Romeo de Freitas, Olivier Missa, Wiltshire Hinds, Julian Pillay, Iwan Derveld, Wilmer Díaz, Corletta Toney, Peter Hoke. Middle (L to R): Jim Sanderson, Marijem Djosetro, Burton Lim, Zacharias Norman, Deirdre Jafferally, Aiesha Williams, Leroy Ignacio, Kathleen Fredericks. Front (L to R): Jacky Sutrisno Mitro, Jan Hendrik Meijer, Lindsford La Goudou, Jensen Montambault.

Recording mammal tracks like this jaguar (*Panthera onca*) print, is one of the methods RAP mammalogists use to get a sense of the faunal community in a short period of time.

First RAP camp, located on the Lower Kwitaro River at the mouth of Pobawau Creek. The habitat in this area included dry and seasonally flooded forest, vine tangles, river edges, and upland forest.

Located about two hours up the Rewa River from the Kwitaro River mouth, this remote area is entirely uninhabited upstream. Carvings, like the one pictured, show that indigenous groups once frequented this now extremely isolated river area.

The Collared Puffbird (*Bucco capensis*) is part of an insect feeding family known for its high whistle, cryptic plumage, and sluggish behavior. This individual has "puffed" its fine feathers around its head in a stress response.

This orchid, *Prosthechea vespa* (Vell.) W.E. Higgins, was found in an isolated mountain peak cloud forest habitat at the Cacique Mountain site.

This female Black-headed Antbird (*Percnostola rufifrons*) represents one of the forest understory species that are ubiquitous in the Guianas. Some species in the families Thamnophilidae (Typical Antbirds) and Formicariidae (Ground Antbirds) are known to eat the insects turned up in the wake of army ants' destructive path, giving the antbird its name.

Practocephalus hemiliopterus is considered to be the most colorful of the large Amazon catfishes. This species feeds on smaller fish, crabs, and palm fruits.

Camera trapping

Large mammals are difficult to see on RAP expeditions, in part because they are nocturnal, have large ranges, and do not call in the manner of birds and frogs. RAP scientists rely on methods such as observing tracks, using live traps, and setting CamTrakker™ remote sensing camera photo-traps to provide other concrete evidence.

This margay (*Leopardus wiedi*) was "captured" on film by a camera trap during the 2001 RAP survey of the Eastern Kanuku Mountains near the Lower Kwitaro River.

This ocelot (*Leopardus pardalis*) was attracted to the strongly scented bait left by the non-volant mammal team and was photographed by a strategically placed camera trap.

Executive Summary

INTRODUCTION

The Kanuku Mountain region is part of a vast pristine wilderness area that stretches across the Guayana Shield. The mountains are approximately 100 km east-to-west and 50 km north-to-south and are divided by the Rupununi River into eastern and western ranges with peaks of up to 1,000 m. The Kwitaro and Rewa rivers of our study area join the Rupununi River to form part of the larger Essequibo River watershed. The surrounding savannah region experiences periodic flooding which, at times, physically joins with Brazil's Branco River savannah, creating a highly diverse flora and fauna with both Guayana Shield and Amazon Basin elements. The geology consists of old, Precambrian rock and largely infertile soils.

Naturalists have visited the Kanuku Mountains for over 150 years, but not until recently has a systematic study of the biodiversity been undertaken. The savannahs have been classified as herbaceous, shrub, open woodland, and closed woodland. The more diverse moist forest ecotone has been divided into habitats including riparian swamp and gallery forests, as well as hill and cloud forests, each with many microhabitats. The Kanukus register a significant percentage of the Guyanese vertebrate diversity, including many species endemic to the Guayana Shield and healthy populations of other species that are rare or endangered elsewhere in the world.

Although there is a very low permanent human presence in the region, potential and actual human-associated environmental pressures do exist. These include large and small-scale mining operations, timber harvest, hunting for food, and wildlife trade as well as traditional land uses. The effects of these pressures are expected to increase substantially upon the completion of the Takutu River bridge at Lethem that will open the region to Brazil.

As a result of this survey, we strongly recommend that the Kanuku Mountains be placed under some degree of legal protection. A protected area and restrictions should be defined in collaboration with local communities and be accompanied by education and outreach programs as well as additional scientific studies.

STUDY AREA

The entire team of RAP scientists surveyed two sites, Pobawau Creek (3°16'3.1"N, 58°46'42.7"W) between 20–24 September and Cacique Mountain (3°11'29.5"N, 58°48'42.0"W) between 25–29 September 2001. Both of these areas were located along the Kwitaro River in seasonally flooded tropical rainforest. One peak of the Kanuku Mountains at 450 m was surveyed at the Cacique Mountain site. The fishes and water quality team sampled in an additional location at the Corona Falls rapids (3°11'35.0"N, 58°48'39.6"W) on the Rewa River two hours upstream from the mouth of the Kwitaro. More detailed sites descriptions can be found in the Gazetteer.

SUMMARY OF RESULTS

Plants

The Kanuku Mountains and surrounding area are known to contain 1,577 plant species, representing 26.3% of Guyana's flora. This RAP survey's preliminary results document 215 plant species, from 198 specimens and more observations, comprising 72 families (Appendix 2). Among these, 40 species appear to be new records for the Kanuku Mountains, and further study and comparison are recommended. The vegetation in the study area was composed mainly of mixed forests on slopes, in valleys, and flood plains. The most common species were *Licania* sp., *Carapa guianensis*, *Eschweilera* spp., *Pouteria* spp., *Catostemma fragans*, *Aspidosperma* sp., *Trattinickia* sp., *Sagotia racemosa*, *Duguetia* spp., and *Rinorea* spp. The abundance of *Bertholletia excelsa* in one of the forests sampled may be due to past human activity. No logging or clearings were observed. The only visible trace of human disturbance was in the form of Balata tree bark (*Manilkara* sp.) cut to collect latex.

Birds

Almost unstudied before the 1993 RAP survey of the Western Kanukus, the avifauna of the Kanuku Mountain region is one of the most diverse in Guyana. This report combines the findings of the 2001 RAP expedition with other surveys carried out in the same region of the Lower Kwitaro River by Davis Finch in 1998 and 2001 for a total of 264 species, bringing the total number of bird species in the Kanuku region to 419 species (or 53% of the species known for Guyana). The cumulative surveys include 63 new records for the region and document 39 species considered uncommon in the Neotropics and two species considered rare, the Harpy Eagle (*Harpia harpyja*) and the Orange-breasted Falcon (*Falco deiroleucus*). As a result of these surveys, we now know that the Kanuku Mountains contain at least 17 of the 25 bird species known to be endemic to the Guayana Shield, reinforcing the need to protect this area. More avifaunal studies, especially in the hill and montane habitats are recommended.

Fishes and Water Quality

The RAP team recorded fishes from the Kwitaro and Rewa rivers within the larger Rupununi basin. The team recorded at least 113 species of freshwater fishes including 45 large-sized food fishes (>15 cm Standard Length) among which the arapaima (*Arapaima gigas*) and arawana (*Osteoglossum bicirrhosum*) are especially noteworthy for their economic value. Because the results of the 2001 RAP survey represent the first inventory of fishes and water quality for the Kanuku Mountain region, they cannot be considered definitive and more extensive studies, especially a low-water survey for smaller fish, are strongly recommended. Even higher fish fauna diversity is expected because the Guayana Shield is known for its high freshwater biodiversity and the Kanukus

are situated where the waters from the larger Amazon and Essequibo catchments occasionally mix.

The waters of the Rewa and Kwitaro rivers exhibited the characteristics of an Amazonian clear-water stream. The Rewa River had a slightly lower conductivity and a higher transparency than the Kwitaro River. Small tributary rainforest creeks in the study area were more acidic than the large rivers.

Mammals

A total of 52 mammal species were identified during the survey at the two sites. The mammal surveys used a combination of live, folding box-style traps, mist-nets, and indirect techniques (photo-trapping, visual sightings, footprints, audible calls). The 125 photos taken by camera photo-traps during the study period recorded 14 species including the large cats, ocelot (*Leopardus pardalis*) and margay (*L. wiedi*). All eight primates known to have ranges in Guyana were found or confirmed to have been observed in this area. The endangered giant river otter (*Pteronura brasilensis*) was observed and photographed and a jaguar (*Panthera onca*) was heard. The 27 bat species recorded on this expedition brings the total bat diversity for the Eastern Kanuku Mountains to an impressive 89 species. With data from previous surveys, the total mammal diversity of the Kanukus is raised to 155 species, representing approximately 70% of the Guyanese mammal fauna. By conserving this region, a large segment of the mammal biodiversity will be protected.

CONSERVATION ELEMENTS

High Species Richness

The results of this RAP survey show that the Kanuku Mountains have an exceptionally high species richness, especially for mammals and birds, which are the best-known taxonomic groups. If the Kanuku Mountains are incorporated into a protected area system that includes both forests and savannahs and representatives of the many microhabitat variations, a high proportion of the plant, bird, fish, and mammal species of Guyana would be protected. The area, especially along the Lower Kwitaro and Rewa Rivers, is extremely isolated from human habitation and current populations of most species appear healthy, indicating that the protected area would be successful and achieve the goals of protecting biodiversity. Immediate environmental pressures to the area are low. Certain risks to environmental stability, especially hunting animals for the wildlife trade (e.g., Ara parrots) and food (e.g., the endangered giant arapaima), however, should be carefully monitored and educational programs should be continued and expanded. No exotic species were found during the RAP survey, indicating a pristine habitat, and steps should be taken to prevent the future introduction of such species.

The Role of Local Communities

Land rights, especially with respect to indigenous peoples and protected areas, are an extremely sensitive issue throughout South America. While it is critical to place the Kanuku Mountain region under some form of legal protection as soon as possible, the success of this protected area depends on open discussion and close collaboration with the communities in the region. These dialogues will facilitate mutual understanding between the government, conservation groups, and the communities. They may reduce potential problems such as poaching or a feeling of community displacement, and emphasize the potential benefits of achieving both conservation and developmental objectives through the establishment of a protected area.

During this expedition, RAP scientists noted that community members were enthusiastic about being involved in scientific surveys, not only as logistic support and field guides, but also as trained field assistants. Several of the participating scientists expressed their interest in continuing to train and work with community members to develop better baseline knowledge and a long-term biological monitoring program in the region.

Participating in scientific training could also prepare local community members to become part of the developing ecotourism market. With a largely undisturbed environment and many charismatic species such as caimans, harpy eagles, large cats, macaws, and giant otters, the region has great potential for ecotourism, but is relatively unpublicized. Truly sustainable ecotourism projects can be arranged to benefit a wide sector of the community and have minimal impact on the environment.

CONSERVATION RECOMMENDATIONS

Conservation Activities

The results of the RAP surveys in the Kanuku Mountains reinforce the need for Conservation International–Guyana and other groups to continue taking appropriate steps to design and establish a protected area in the Kanuku Mountain region, and to encourage the management of the area in a sustainable manner to achieve both conservation and development for local communities. Specific recommendations from the RAP team follow.

Kanuku Mountain Protected Area

The conservation community should continue working toward placing the Kanuku Mountains under some status of legal protection especially considering that the current level of environmental pressures is low and may be affected once the Takutu bridge at Lethem is completed. Protecting this area will conserve a large portion of the plant, mammal, bird, and fish species known to exist in Guyana, depending on the exact delineation of boundaries and establishment of a sound management program. The protected area must incorpo-rate both of the major habitat types in the region including forest and savannah. In addition, a variety of the microhabitats present such as gallery forest, swamps, and montane areas should also be included. We encourage seeking funds to establish a ranger-training program composed of local inhabitants from the surrounding areas and a field station similar to what has been implemented in Iwokrama Forest in Central Guyana. After being trained in biodiversity assessment and monitoring techniques, these rangers would be responsible for a biological inventory of the protected area and a monitoring program to track the status of biodiversity and the environment. In particular, attention should focus on the management of species susceptible to over-hunting. This can be accomplished through long-term partnerships between international institutions and local groups including the University of Guyana and communities in the region.

Species Protection

We recommend continuing and expanding the existing educational outreach programs to indigenous communities in the Kanuku Mountains. These programs should promote understanding of the need to conserve birds in the region commonly captured for the wildlife trade (e.g., parrots, birds of prey) and sustainable hunting practices that can help to prevent local extinctions. We also recommend working with these communities to heighten awareness of existing legislation on the endangered arapaima, and its importance for immediate conservation.

Ecotourism Promotion

Inhabitants of Shea, Maruranau, and Awariwaunau villages in the south and Rewa, Apotari, and Annai villages in the north utilize the Eastern Kanuku Mountains for hunting, fishing, and harvesting non-timber forest products. Duane de Freitas of Dadanawa Ranch operates an ecotourism business that organizes an average of two trips per year to the Kwitaro and Rewa rivers, among others. We believe these activities should continue if protected area status is granted to the Kanuku Mountains and indeed argue that such activities are integral to maintaining the biodiversity in this vast, nearly pristine area.

The Kanuku Mountain ecosystems should be promoted to potential ecotourists in the United States and Europe. Combining research-based opportunities with education-based ecotourism, for instance, can generate revenue. Bird and primate observation will probably continue to be the main attraction. However, other alternative nature activities such as nocturnal bat netting can be easily started with minimal training, providing alternative livelihoods for people in and around the area.

RECOMMENDED SCIENTIFIC STUDIES

See individual chapters for more details.

Plants

The plant communities on top of mountains, slopes, and intermountain valleys should be surveyed for a better understanding of their floristic richness, habitats, and endemism, especially on peaks over 500 m high. A more complete comparison of the flora of the Eastern and Western Kanuku ranges in the context of known Guayana Shield vegetation communities should also be explored.

Fishes

A second fish-sampling expedition during the low-water season (January–February) should be conducted to obtain a more complete and representative list of the fish biodiversity of the Eastern Kanuku Mountains, especially with respect to small-sized species. Studies on the biology, ecology, and culture of the world's largest freshwater fish, the endangered *Arapaima gigas* (also a flagship species of the Kanuku Mountain area), should also be initiated.

Birds

Small cryptic birds in the under-story should be surveyed with mist-nets, and the habitat preferences of uncommon or endemic birds, which potentially require the most urgent protection, should also be studied. Populations of potentially threatened bird species such as the Harpy Eagle (*Harpia harpyja*) which tends to disappear throughout its range, parrots that may become threatened by the wildlife trade, and game birds should be monitored.

Non-Volant Mammals

We strongly recommend that the camera-trapping monitoring protocol be incorporated into community programs involving local Amerindian villages including Shea, Nappi, and Shulinab located near the Kanuku Mountains. A sustained long-term camera photo-trapping research and outreach program can be undertaken to provide villagers and scientists with continuous information on local wildlife resources, thus enabling more informed conservation actions.

Small Mammals

The current state of biological knowledge for mammal diversity in the Kanuku Mountain region is relatively good compared to other faunal groups, primarily because of the earlier long-term involvement of Dadanawa Ranch collecting in the savannahs and adjacent forest in the south Rupununi. Despite this sound taxonomic base, in only eight days we collected five species of small mammals previous unknown from the eastern region. This indicates that more biodiversity survey work is needed to document the actual diversity, abundance, and local distribution of mammals that live in this area.

Herpetofauna

Additional participants intended to inventory amphibians and reptiles were prevented from traveling to the RAP site due to restrictions following the September 11th terrorist attacks on the United States. Therefore, the herpetofauna of the Eastern Kanuku Mountain region remains relatively unstudied. However, the area is known to support healthy populations of species considered at risk in other parts of the world including the black caiman (*Melanosuchus niger*) and giant river turtle (*Podocnemis expansa*). These factors warrant carrying out a rapid survey of the herpetofauna in the Eastern Kanuku Mountains as soon as possible.

Chapter 1

An Ecological, Socio-economic, and Conservation Overview of the Kanuku Mountain Region, Southern Guyana

Olivier Missa and Jensen R. Montambault

ABSTRACT

The Kanuku Mountain region, located in southwestern Guyana, is predominately covered by forest habitat (ranging from cloud forest to swamp land) and is almost completely surrounded by the expansive Rupununi savannah. Scientific knowledge of this area is extremely limited; the 1993 Rapid Assessment Program (RAP) expedition to the Western Kanukus strongly recommended studying the Eastern Kanukus. This zone is notable for healthy populations of many large vertebrates that are endangered or locally extinct in other regions of the world, as well as for housing a high percentage of the known Guyanese fauna. Environmental pressures such as mining and timber extraction are currently minimal, but are expected to increase upon the completion of the bridge over the Takutu River at Lethem, which will open up access to and from Brazil. Hunting, wildlife trade, and traditional land uses do have some current environmental impact. There is substantial biological evidence to support placing the Kanukus under some form of legal protection.

ECOLOGICAL OVERVIEW

Geography

The Kanuku Mountain range is located in the Rupununi region of southwestern Guyana and is separated into the Western and Eastern Kanukus by the north-south course of the Rupununi River. The slopes of these mountains are largely covered by closed-canopy forests and they are surrounded by the Rupununi savannah, part of Guyana's largest expanse of grasslands (LOC, 1992). At 120–150 m elevation, these grasslands are ecologically connected to Brazil's Branco River savannah. Toward the east, however, the Kanukus join the vast pristine tropical forest expanse of the Guayana Shield shared with neighboring Brazil, Suriname, Venezuela, and also French Guiana and parts of Colombia (CEP, 1999; Mittermeier et al., 1998). Extending 100 km east-to-west and 50 km north-to-south, the Kanukus form an impressive landscape feature. The western range rises steeply to its highest peak at 1,067 m with several minor peaks over 900 m above sea level, whereas the eastern range has a more gentle slope and greater expanse, with the highest peak measuring 900 m and an overall average of 450 m (LOC, 1992). In addition to the Rupununi River, the other major waterways include the Kwitaro River, flowing northward along the eastern edge of the Kanukus to join the Rewa River above its mouth in the lower, eastern portion of the Rupununi (Parker et al., 1993; Map).

Geology

The Kanuku Mountains lie in the Precambrian sediments of one of the oldest rock formations on earth (over 2 billion years old). The rocks in this region include a metamorphic component, either schist or gneiss mixed with igneous intrusions like granite, dolerite, or diorite. The mountains are made of the more resistant basic and metamorphic volcanic rocks, whereas the more fragile schists and gneissose granites form the surrounding savannahs (Agriconsulting, 1993; Boggan et al., 1997).

Soils

Several types of soils, mostly infertile, are found in the Rupununi region (Agriconsulting, 1993). The two most common types in the Kanuku Mountains are the acidic lithosols on ridges and steep hills, as well as the reddish-brown latosols on gently sloping terrain and hill pediments of the northern side of the Western Kanukus. Poorly drained low-humic gleys and ground water laterites are found in the alluvial plains. Elsewhere, red-yellow latosols and regosols predominate. Overall, these soils lack important nutrients like potassium, phosphorus, sodium, magnesium, and calcium, and are heavy in aluminum, which may become toxic to plants.

Hydrological System

Three main rivers drain the region. The Takutu River in the western savannah flows into the Branco River in Brazil, joining the Amazon Basin. The Rupununi River has a northerly course cutting through the two Kanuku ranges to the foot of the Pakaraima Mountains, then continues eastward to join the Essequibo River (the largest basin in Guyana). The Kwitaro River flows north along the eastern edge of the Kanuku Mountains, and then joins the Rewa River before it merges with the Lower Rupununi. The headwaters of the Takutu and Rupununi rivers in southern Guyana and Brazil join occasionally in years of extreme flooding, allowing the mixing of fish communities normally segregated into either Amazon or Guayana Shield river systems.

The northern stretch of the Rupununi River and, to some extent, the Lower Rewa are prone to yearly flooding cycles. During the dry season, most small creeks dry out completely and the lower river water levels expose rocks and rapids, making them more difficult to navigate (Parker et al., 1993). Ponds and lakes subsist for several months in areas where clay soil prevents drainage. Vertical fluctuations of three to nine meters are common between the water levels of the rainy and dry seasons (Agriconsulting, 1993).

Climate

The wet season in Southern Guyana is from May to August (average monthly rainfall above 100 mm) and the dry season is from September to April (monthly rainfall below 100 mm), in contrast to the rest of the country, which has two wet and two dry seasons. Annual rainfall is between 1500–2000 mm per year with the heaviest rain in May. The mean annual daily temperature is 27.5° C and fluctuates little throughout the year. During the dry season, rainfall is caused primarily by thunderstorms, especially in December (LOC, 1992).

Scientific Explorations

Although naturalists have visited the Kanuku Mountains since the mid-1800s (Parker et al., 1993), this region is still considered relatively poorly known. While plants, birds, and mammals have been studied in this region, biological surveys continually add to the species lists and knowledge about habitat use.

The most significant vegetation studies came from the botanical explorations conducted by Utrecht University (ter Welle et al., 1987, 1990, 2000), the Biological Diversity of the Guianas program of the Smithsonian Institution (Parker et al., 1993; Boggan et al., 1997), and from the forestry industries development surveys of the Food and Agriculture Organization of the United Nations carried out in southern Guyana (FAO, 1970).

Knowledge of mammal populations began slowly in 1900 by J.J. Quelch's collection of 28 species (Thomas, 1901), then intensified in the mid 1960's, when S.E. Brock collected a large number of specimens around Dadanawa Ranch and along the Kwitaro and Rewa Rivers for the Royal Ontario Museum and the United States National Museum (Emmons, 1993; de Freitas, pers. comm.). The Royal Ontario Museum has since made several collecting trips in the Kanuku Mountain region (Lim and Norman, 2002).

The avifauna in the Kanukus has been highlighted in ecological and behavioral studies of two remarkable species: the Guianan Cock-of-the-Rock (*Rupicola rupicola*; Gilliard, 1962) and the Harpy Eagle (*Harpia harpyja*; Rettig, 1977, 1978, 1995). Although the region has attracted the attention of bird enthusiasts, a list of bird species present specifically in the Kanuku Mountain region was only recently published (Parker et al., 1993). In addition, Davis Finch of WINGS (Finch et al., 2002), Mark Robbins of the University of Kansas, Michael Braun and others affiliated with the Smithsonian Institution have conducted bird studies in this area over the last decade (Braun et al., 2000).

In 1993, a team of scientists organized by Conservation International's Rapid Assessment Program (RAP) assessed the biodiversity of the Western Kanuku Mountains and Rewa River for the first time, using key taxonomic groups: plants, mammals, birds, reptiles, amphibians, and coprophagous beetles. Their report (Parker et al., 1993) remains an excellent source of information available on the biodiversity of this incredibly rich and remote area in Guyana. The combination of these results and museum and literature records suggested that the Eastern Kanuku Mountains would be even richer than their Western neighbors.

Vegetation and Habitat Diversity

Two main ecotones exist in the Rupununi region, savannah and moist forest. These savannahs occur primarily because of soil rather than climatic conditions. The hard clay layer underlying the highly weathered white sands is difficult for tree roots to penetrate and limits their access to ground water during the dry season. Thus, moist forest only occurs on more porous substrates such as in hills on granite outcrops and latosols, along rivers as gallery forest in alluvial sands, and in scattered "islands" where the clay layer has been penetrated (Clarke et al., 2001). During this RAP survey we recognized four broad categories of savannahs. The *herba-*

ceous savannah is dominated by grasses and sedges and is distinct from the *shrub savannah* that is abundant in the seasonally flooded Upper Rupununi. *Open savannah woodland* and *closed savannah woodland* occur on the higher ground of the Southern Rupununi, where Leguminosae was the most abundant tree family and Cyperaceae the most abundant herb (Agriconsulting, 1993).

The moist forest ecotone contains a wider diversity of habitats including *marsh* and *gallery forest* along rivers, *mixed forest* on the hills, and *cloud forest* on the mountaintops. Although some vegetation patterns are beginning to emerge from the data already collected (Parker et al., 1993; ter Steege, 2001; Jansen-Jacobs and ter Steege, 2000), there is still much work needed in southwestern Guyana to characterize and map plant communities.

Marsh forest develops on seasonally flooded terrain along the lower portions of the Rupununi and Rewa rivers (Foster, 1993; ter Steege, 2001). This forest is referred to as "Mora forest" and is dominated by the *Mora excelsa* tree. In many cases, up to 80% of the trees above 30 cm in diameter belong to this single species. Other common tree species include *Carapa guianensis* (Crabwood) and *Pterocarpus officinalis*. In the savannah region, Mora forest is replaced by *gallery forest* along the rivers, with a mix of tree species, the most common of which are *Caryocar microcarpum*, *Macrolobium acaciifolium*, *Senna latifolia*, *Zygia cataractae*, and *Genipa spruceana* (ter Steege, 2001). Along the Rupununi River, where soil conditions are dryer, there exists instead a *riverine scrub* habitat dominated by small trees and scrub (Agriconsulting, 1993).

The Kanuku Mountains contain a narrow band of *dry deciduous forest* of mixed composition immediately adjacent to the savannah. Common tree species include *Goupia glabra*, *Couratari* sp., *Sclerolobium* sp., *Parinari* sp., *Apeiba* sp., *Peltogyne* sp., *Catostemma* sp., *Spondias mombin*, and *Anacardium giganteum* (ter Steege, 2001).

The forest on the foothills and slopes of the Kanuku Mountains is 25 m high with a few emergent trees reaching 40 m. It is classified as a *mixed semi-deciduous dry forest* and differs substantially in its composition from the southern Guyana rainforests (Jansen-Jacobs and ter Steege, 2000). It is characterized by *Cordia alliodora*, *Centrolobium paraense*, *Apeiba schomburgkii*, *Acacia polyphylla*, *Pithecellobium* sp., *Peltogyne pubescens*, *Manilkara* sp., *Cassia multijuga*, and *Vitex* sp. (ter Steege, 2001).

The highest peaks of the Kanuku Mountains contain 5–10 m tall *cloud forest* trees such as *Couepia canomensis*, *Stelestylis stylaris*, *Sphyrospermum cordifolium*, *Rhodostemonodaphne scandens*, *Marcgraviastrum pendulum*, *Cybianthus detergens*, and *Cybianthus roraimae* (Jansen-Jacobs and ter Steege, 2000). These ecosystems are thought to house unique communities and warrant further study, particularly in the Eastern Kanukus (Díaz, 2002).

The southern Guyana forest differs from the central and northwestern regions in part due to the presence of the following genera: *Anacardium*, *Andira*, *Bagassa*, *Cecropia*, *Couratari*, *Dipteryx*, *Geissospermum*, *Laetia*, *Micropholis*, *Parkia*, *Pourouma*, *Pseudopiptadenia*, *Qualea*, *Sclerolobium*, *Simarouba*, *Tetragastris*, *Virola*, and *Vochysia* (ter Steege and Zondervan, 2000). At the local level, the forests of southern Guyana also have higher tree alpha-diversity than the others (ter Steege, 1998), which tend to be strongly dominated by a few species (Fanshawe, 1952; ter Steege et al., 1993). At the regional level, however, ter Steege et al. (2000) recently suggest that southern Guyana has lower diversity than other regions of Guyana, especially the Pakaraima Highlands and central region. Although it may be true that the regions have different levels of habitat heterogeneity, the plant families chosen for analysis may have also underestimated the overall diversity of the dryer southern Guyana. Clusiaceae, Myrtaceae, Bombacaceae, Anacardiaceae, and especially Leguminosae were all absent from the analysis and dominate the species richness of southern Guyana vegetation. Forests in this region have a relatively low number of species endemic to Guyana and typical Guyanese plants are replaced by Amazonian elements in a southerly progression.

Fauna

The animal species recorded in Guyana include 786 birds (Braun et al., 2000), 225 mammals (Engstrom and Lim, 2000) including 121 bats (Lim and Engstrom, 2001), 119 amphibians (Reynolds et al., 2001), 153 reptiles (Reynolds et al., 2001), and 690 fishes (Lasso, 2002).

For comparison, the number of species recorded from the Kanuku Mountains to the date of this publication (including our findings) is 419 birds (over 50% of all Guyanese avifauna), 155 mammals including 89 bats (approximately 70% of Guyana's mammal fauna), 20 amphibians, 23 reptiles, and 113 fishes, while the fishes represent 16% of the freshwater fish species known from Guyana, it should be noted that our preliminary results were the only surveys ever carried out in this region, and should not be considered complete. Very little is known about other faunal groups in this region, particularly invertebrates.

The Kanuku Mountains are also remarkable for harboring healthy populations of many species which are threatened elsewhere. These include, but are not limited to, the giant river otter (*Pteronura brasiliensis*), giant anteater (*Myrmecophaga tridactyla*), giant armadillo (*Priodontes maximus*), giant river turtle (*Podocnemis expansa*), harpy eagle (*Harpia harpyja*), black caiman (*Melanosuchus niger*), and arapaima (*Arapaima gigas*)–the largest Neotropical freshwater fish (Parker et al., 1993).

Indigenous population

The 15,000 indigenous inhabitants (referred to as Amerindians) of southwestern Guyana are divided into two main groups: the Wapisiana, 6,000 people of Arawak origin living in the Southern Rupununi, and the Macusi people of Carib decent, numbering 9,000 in the Northern Rupununi (LOC, 1992). Both groups live primarily on the savannahs surrounding the Kanukus, practicing agriculture based on crop rotation in the ecotone between savannah and forest, cultivating varieties of cassava, supplemented with yams, sweet potatoes, rice, eddoes, dasheen, and corn as staples and pumpkin, banana, watermelon, papaya, pineapple, and mango as fruits. They also rear some livestock and regularly travel into the Kanuku Mountains to hunt, fish, and harvest forest products (Forte, 1990).

Balata latex harvesting from *Manilkara* sp. trees was a major economic activity for the Amerindian people of this region as recently as the 1970's, but has declined sharply with the introduction of similar synthetic materials to the world market. Generally, economic opportunities for Amerindians in this area are limited to subsistence agriculture, capturing animals for the pet trade (Edwards, 1992), and small-scale gold and diamond mining.

LEGAL PROTECTION STATUS AND ENVIRONMENTAL PRESSURES

The Kanuku Mountains are presently under no legal protection status, despite several biological studies stressing the importance of conserving the area (e.g., Agriconsulting, 1993; Parker et al., 1993). Critical as protecting this area is, it is equally important to consider the local communities' needs, rights, and expectations for an effective long-term conservation project in the Kanukus (NDS, 1996).

The current pressures on the environmental stability of the Kanuku Mountain region of southern Guyana are minimal. They are expected to increase once the bridge over the Takutu River at Lethem, currently under construction, is completed linking southern Guyana and Brazil (Stabroek News, 2001). Four specific activities with the potential to disturb natural habitats in the Kanukus are mining, timber extraction, hunting, and agricultural land uses.

Mining

Three main minerals are economically exploited in Guyana: bauxite, gold, and diamonds, in order of export value. Official gold and bauxite production combined account for nearly one third of the national export, or US$160 million annually. Major bauxite mines are controlled by the state run Bauxite Industry Development Company and do not affect our study area. The largest gold producer, Omai Gold Mines, Ltd., is controlled by Canadian Cambor, Inc. with other investors in the US and Guyana (Gurmendi, 1997); its operations are limited to the Essequibo River Basin, which also does not affect our study area. Vanessa Ventures, Ltd.'s subsidiary, Romanex, has performed exploratory gold and diamond mining operations on the Kwitaro River, upstream from our sites at Marudi Ridge and Mazoa Hill, and has recently announced that it will begin to mine the surface gold resources in this area (NewsWire, 2002). Illegal gold miners from Brazil, known as *garimpeiros*, are becoming increasingly active on the southern border with Brazil and are sometimes subsidized by large-scale operations. In the

Third International Conference on Environmental Enforcement, the Guyanese government stated, "an examination of various mining activities in Guyana… shows that, invariably, mining destroys valuable stands of forests and wildlife habitat" (Singh, 1994). Risks include pollution in the form of siltation and heavy metals, increased road development, and hunting by miners and their families. There is little information available on the environmental impact of operations in our study area.

Timber Extraction

Over half the forests in Guyana (52%) lie within State Forest boundaries, of which a quarter were allocated for harvesting in 1996. In 1997, timber exports generated approximately US$15 million in revenue, not including an additional US$21 million from plywood exports (GFC, 1998). Plywood prices, however, are below production costs and show little prospect of being profitable in Guyana.

Of the species exported, 80% are dense hardwoods, especially the endemic greenheart (*Chlorocardium rodiei*), which alone represents 40% of the total round-wood production. The overseas market has not been profitable for trees other than greenheart, so the other species are exploited primarily for a domestic market. Greenheart is most abundant in central Guyana, in particular on brown sand soil (ter Steege et al., 1993), and is scarce in southern Guyana (FAO, 1970). In addition, southern Guyana's forests are virtually inaccessible, with few roads or navigable rivers, and low soil fertility makes it impossible to convert deforested areas to oil palm plantation to offset a poor timber harvest. These factors indicate that the Kanuku Mountain and Lower Kwitaro River forests are unlikely to be exploited for timber on a large scale in the near future.

Hunting and Wildlife Trade

Overall, the legal wildlife trade in Guyana is valued at US$800,000 annually, and the country is the fifth largest exporter of birds in the world (NDS, 1996). Live birds trapped for export include, among many others, the Orange-winged Amazon (*Amazona amazonica*), Yellow-crowned Amazon (*Amazona ochrocephala*), and the Blue and Yellow Macaw (*Ara ararauna*). In southwest Guyana (Rupununi region), this trapping is primarily carried out by Amerindian communities and represents a significant part of their cash economy (Edwards, 1992). Although little concrete information exists on how the legal and illegal trade impacts wild populations, it has been suggested that the once common Sun Parakeet (*Aratinga solstitialis*), is locally extinct due to hunting pressure (Parker et al., 1993). Due to its proximity to Brazil, the Kanuku Mountain region faces a larger risk than other areas of southern Guyana of populations being decimated by the wildlife trade. Management of the wildlife trade in Guyana has been described as "ad hoc," primarily due to poor infrastructure in government and enforcement agencies and a severe lack of biological data necessary to make concrete recommendations (Singh, 1994).

Studies of the Guayana Shield and Amazon Basin indicate that subsistence hunting poses additional pressures to large vertebrate populations (Peres and Dolman, 2000). In the Eastern Kanuku and Lower Kwitaro region, large mammals such as agoutis, deer, and peccaries are hunted for food (Sanderson and Ignacio, 2002). The many species of large fish present in the rivers do not appear to be over-fished, with the possible exception of the endangered giant arapaima (*Arapaima gigas*), the largest freshwater fish in the Neotropics, which is also illegally captured live for export to Brazil (Mol, 2002). Jaguars, pumas, and other large carnivores are systematically hunted in the savannahs and nearby forests to protect livestock (de Freitas, pers. comm.).

Traditional Land Uses

The Amerindians in the Kanuku Mountain region practice crop-rotation agriculture in the transition zone forest and savannah habitat. Although this practice has been sustainable for centuries, the land they have available has been reduced, and so can no longer be left fallow for sufficient time to recover its fertility (Singh, 1994). In addition, scrubland and savannahs are burned over, leaving grazing land which will lose its viability quickly (Singh, 1994). Wildfires also frequently rage out of control and infringe on the forest-line of the Kanuku Mountains on an annual basis (de Freitas, pers. comm.).

ENVIRONMENTAL CONSERVATION IN THE KANUKU MOUNTAIN REGION

Conservation Importance

The Guayana Shield is classified as one of three Tropical Wilderness Areas that still contain large tracts of land in "pristine" ecological condition (Mittermeier et al., 1998). This vast region of 1,800,000 km² comprises parts of Venezuela, Colombia, and Brazil, and the entirety of Guyana, Suriname, and French Guiana. The Guayana Shield supports a high number of endemic plants and animals. It is estimated that 138 tree genera (Berry et al., 1995) and 40% of the 10,000+ plant species (Boggan et al., 1997) are endemic to this region, and it is thought that no other region harbors more herbaceous systems (Mittermeier et al., 1997).

Within the country of Guyana, tropical forests cover 80%, or approximately 170,000 km² of the land area—70% of which is considered to be in pristine condition. Guyana also boasts one of the lowest human population densities in South America, with about 800,000 people (less than 5 people/km²). Fewer than 50,000 of these inhabitants live in the interior, primarily in indigenous communities (Forte, 1990), while the remainder are concentrated in the fertile agriculture zone along the coast. These demographics and the relatively unexploited natural resources place Guyana in the almost unique position of being able to proactively manage its resources before increasing pressures from logging and mining and other human developments change the status

quo (Hilty, 1982). Guyana is currently the only country in the Western Hemisphere without a national protected area system, and implementing such a system is critical to maintaining its vast biological resources.

The Government of Guyana is already committed to developing a national protected area system, but requires more biological information upon which to base its decisions (Maynell, 2001). Kaieteur National Park is the oldest protected area, legally established in 1929 for the scenic beauty of the longest single-drop waterfall in the world located within its boundaries (CEP, 1999). Iwokrama Forest has now been placed under protection as well, with about 50% of the land area designated as a wilderness preserve. Several other areas, including the Kanuku Mountains and the New River Triangle (Funk et al., 1999; ter Steege et al., 2000) have been proposed as protected areas and await formal recognition. Although the flora and fauna of Guyana are relatively well known at the country level, detailed information on their specific distribution is rarely available. This may bias analyses and conclusions drawn when attempting to establish conservation priorities among potential protected areas (Funk et al., 1999; ter Steege et al., 2000), and was an impetus for a Priority Setting Workshop for the Guianas, facilitated by Conservation International, the Netherlands Committee of the World Conservation Union (NC-IUCN), and the United Nations Development Programme (UNDP) in 2002 (see Conservation Activities section; CABS, 2002).

The Kanuku Mountains are considered important for conservation, based on (1) high vertebrate diversity, in particular birds and mammals (Parker et al., 1993), (2) the presence of healthy populations of many species which are threatened in other part of the world such as the harpy eagle (*Harpia harpyja*), giant river otter (*Pteronura brasiliensis*), giant armadillo (*Priodontes maximus*), giant anteater (*Myrmecophaga tridactyla*), giant arapaima (*Arapaima gigas*), black caiman (*Melanosuchus niger*), and giant river turtle (*Podocnemis expansa*) (Parker et al., 1993), (3) the wide range of habitats, from the swamp forest along the Rupununi River to the cloud forest on the mountaintops (Jansen-Jacobs and ter Steege, 2000; ter Steege, 2001), and (4) the high tree species richness in the plant communities, combining Guayana Shield and Amazonian elements (ter Steege, 1998; ter Steege et al., 2000). The area is ideal for environmental conservation given the absence of large human settlements, although indigenous communities exploit the area sporadically for hunting and harvesting and rivers do provide access to the area.

Current Conservation International Activities in the Kanuku Mountain Region

On a national and local level, Conservation International–Guyana (CIG), through its Georgetown and Lethem offices, is implementing several activities in the Rupununi region to support the establishment of the Kanuku Mountains as a protected area. These activities include a Community Resources Evaluation project (CRE), leadership training, and enterprise development.

The Community Resources Evaluation (CRE), with its objective of clarifying the spatial and temporal patterns of resource use in the Kanuku Mountain region, is one of the most important activities being implemented. The purpose of the exercise is to generate important data on where resources are being used by communities to guide the establishment of a suitable protected area. The CREs are also an opportunity to consult with the communities, provide further background information on the possibility of establishing a protected area in the Kanukus, and to discuss or clarify any issues with the communities.

The Captains and members of the Village Council in each stakeholder community in the Kanuku Mountain region are also being trained in conservation concepts and the process of establishing this protected area. The concept behind this leadership training is to provide a foundation for an informed consensus on establishing the proposed Kanuku Mountain protected area. By providing accurate information, potential misconceptions and misinformation about the protected area will be avoided. The leaders, working together with Community Coordinators identified by the individual communities, are then expected to disseminate the information acquired within their communities to stimulate widespread understanding of the process to establish a protected area. It is hoped that this heightened awareness and understanding will also stimulate greater support for this process. In addition, CIG has identified a comprehensive list of stakeholders for the proposed protected area who are regularly kept informed of the protected area status and have an opportunity to provide any appropriate feedback on the process.

Finally, CI-Guyana has been working along with balata artisans for several years to support the development and sustainable marketing of craft products created from the latex of the Bulletwood tree (*Manilkara bidentata*). This support includes providing management, marketing, and computer related training as well as direct assistance with distributing the products to overseas and local markets. A business plan is currently being finalized and the group is being nurtured to reach a point where CI-Guyana's support would no longer be required.

Regional Biological Corridors and Future Rapid Assessment Program Surveys

As part of its strategy to protect biodiversity, Conservation International has suggested establishing a biological corridor that extends from the Western Kanuku Mountains eastward to Suriname. The 1993 and 2001 Rapid Assessment Program (RAP) expeditions provided efficient and reliable biological data to support the conservation strategy for these regions which have been very difficult to access. An additional inventory using RAP methods is planned for a Conservation Concession that Conservation International is seeking to acquire from the government on a contiguous segment of land on the Essequibo River. This innovative procedure would pay the Government of Guyana the same fees as a timber conces-

sionaire. The RAP expeditions are also expected to assist in developing biological monitoring protocols for the concession and protected areas once established.

Soon after this RAP expedition to Guyana, Conservation International (CI), the Netherlands Committee of the World Conservation Union (NC-IUCN), and the United Nations Development Programme (UNDP) also facilitated a multi-national Conservation Priority Setting Workshop for the Guayana Shield from April 5–9, 2002 in Paramaribo, Suriname. Over 120 of the world's leading biologists, including some RAP scientists and CI staff, participated in this workshop. The objectives of this workshop were: i) to bring together biological and social scientists, conservation professionals, and representatives from government agencies and stakeholder groups to determine the state of biodiversity and ecosystem knowledge of the Guayana Shield; ii) to identify critical issues affecting biodiversity, establish consensus on priority conservation areas, and identify both pressures to the biodiversity and opportunities for its conservation; and iii) to provide decision-makers and stakeholders with the best available information on the ecosystem's biological resources, the area's socio-economic conditions, and proposed actions for conservation programs and research (CABS, 2002).

The resulting technical statement was signed by all workshop participants, and will be presented to regional governments and press, as well as by NC-IUCN at the 2002 Rio +10 Earth Summit in Johannesburg, South Africa. In addition, UNDP-Guyana agreed to facilitate an international policy-level network among the countries of the region, and all parties agreed that multinational research teams were needed in order to conduct biological surveys of unknown areas within the Guayana Shield region. The Rapid Assessment Program plans to follow-up on this last recommendation by conducting an AquaRAP (freshwater biodiversity inventory) in Suriname's Coppename River Basin, scheduled for March 2003. The AquaRAP will survey all three of the upper tributaries, the Rechter, Linker, and Midden Rivers and continue down the main channel of the Coppename to the mouth. Fish, plants, invertebrates, and water quality will be assessed in this system that combines black and clear water. These upper reaches of the Coppename, along with its headwaters, have never before been studied, and the information gained through the month-long assessment will be invaluable in the development of a management plan for the Central Suriname Nature Reserve. In addition, the results of the Suriname expedition, this Guyana biodiversity assessment, and the 2000 AquaRAP survey of the Caura River basin in Venezuela will be comparable as baseline biodiversity inventories using RAP methodology in three distinct regions of the Guayana Shield that were previously little known to science.

LITERATURE CITED

Agriconsulting. 1993. Preparatory study for the creation of a protected area in the Kanuku Mountains region of Guyana. Unpublished Report. European Development Fund, Rome, Italy.

Berry, P.E., O. Huber, and B.K. Holst. 1995. Floristic Analysis. *In*, P.E. Berry, B.K. Holst, and K. Yatskievych (eds.). Flora of the Venezuelan Guayana. Vol. 1— Introduction. Missouri Botanical Garden, St Louis, Missouri. Pp. 161–191.

Boggan, J., V. Funk, C. Kelloff, M. Hoffman, G. Cremers, and C. Feuillet. 1997. Checklist of the Plants of the Guianas, 2nd edition. Centre for Biological Diversity, University of Guyana, Georgetown, Guyana.

Braun, M.J., D.W. Finch, M.B. Robbins, and B.K. Schmidt. 2000. A Field Checklist of the Birds of Guyana. Biological Diversity of the Guianas Program Publication 41. Smithsonian Institution, Washington, DC.

CEP (Caribbean Environment Program). 1999. Guyana Country Profile. *In*, Status of Protected Area Systems in the Wider Caribbean Region. CEP Technical Report 36–1996. United Nations Environmental Programme, Kingston, Jamaica, West Indies.

CABS (Center for Applied Biodiversity Science). 2002. Conservation priorities identified for Guayana Shield. Conservation International, Washington, DC.

Clarke, H.D., V.A. Funk, and T. Hollowell. 2001. Using checklists and collection data to investigate plant diversity: A comparative checklist of the Plant Diversity of the Iwokrama Forest. SIDA, Botanical Miscellany 21. Botanical Research Institute of Texas, Fort Worth, Texas. 86 pp.

Díaz, W. 2002. The Vegetation Along the Lower Kwitaro River on the Eastern Edge of the Kanuku Mountain Region, Guyana. *In*, Montambault, J.R. and O. Missa (eds.). 2002. A Biodiversity Assessment of the Eastern Kanuku Mountains, Lower Kwitaro River, Guyana. RAP Bulletin of Biological Assessment 26. Conservation International, Washington, DC.

Edwards, S.R. 1992. Wild Bird Trade: Perceptions and Management in the Cooperative Republic of Guyana. *In*, Thomsen, J.B., S.R. Edwards, and T.A. Mulliken. Perceptions, Conservation, and Management of Wild Birds in Trade. TRAFFIC International, Cambridge, United Kingdom.

Emmons, L.H. 1993. Mammals. *In*, T.A. Parker, III, R.B. Foster, L.H. Emmons, P. Freed, A.B. Forsyth, B. Hoffman, and B.D. Gill (eds.). A biological assessment of the Kanuku Mountain region of southwestern Guyana. RAP Working Papers 5. Pp. 17–24. Conservation International, Washington, DC.

Engstrom, M. and B. Lim. 2000. Checklist of the mammals of Guyana. The Biological Diversity of the Guianas Program, Smithsonian Institution, Washington, DC.

Fanshawe, D.B. 1952. The vegetation of British Guyana: A preliminary review. Imperial Forestry Institute Paper 29. Oxford, United Kingdom.

FAO (Food and Agriculture Organization of the United Nations). 1970. Forest Industries Development Survey, Guyana. Reconnaissance Survey of the Southern Part of Guyana, based on the work of R. de Milde and D. de Groot. FO: SF/GUY 9, Technical Report 15. 65 pp. Rome, Italy.

Finch, D., W. Hinds, J. Sanderson, and O. Missa. 2002. Avifauna of the Eastern Edge of the Eastern Kanuku Mountains, Lower Kwitaro River, Guyana. *In*, Montambault, J.R. and O. Missa (eds.). 2002. A Biodiversity Assessment of the Eastern Kanuku Mountains, Lower Kwitaro River, Guyana. RAP Bulletin of Biological Assessment 26. Conservation International, Washington, DC. Pp. 17–24.

Forte, J. 1990. The populations of Guyanese Amerindian Settlements in the 1980's. Amerindian Research Unit, University of Guyana, Georgetown, Guyana.

Foster, R.B. 1993. Site descriptions and vegetation. *In*, T.A. Parker, III, R.B. Foster, L.H. Emmons, P. Freed, A.B. Forsyth, B. Hoffman, and B.D. Gill (eds.). A biological assessment of the Kanuku Mountain region of southwestern Guyana. RAP Working Papers 5. Conservation International, Washington, DC.

Funk, V., M.F. Zermoglio, and N. Nasir. 1999. Testing the use of specimen collection data and GIS in biodiversity exploration and conservation decision-making in Guyana. Biodiversity and Conservation. 8: 727–751.

Gilliard, E.T. 1962. Strange courtship of the Cock-of-the-Rock. National Geographic. 121(1): 134–140.

Gurmendi, A. 1997. The mineral industry of Guyana. Minerals Information, U.S. Geological Survey, Reston, VA.

GFC (Guyana Forestry Commission). 1998. Forestry in Guyana: Market Summary 1997. Georgetown, Guyana.

Hilty, S.L. 1982. Environmental Profile on Guyana. *In*, World Directory of Country Environmental Studies (1991/93 and diskette eds.) U.S. Agency for International Development, Washington, DC.

Jansen-Jacobs, M. and H. ter Steege. 2000. Southwest Guyana: a complex mosaic of savannahs and forests. *In*, H. ter Steege (ed.). Plant Diversity in Guyana, with recommendations for a National Protected Area Strategy (Tropenbos Series 18). Pp. 147–157. The Tropenbos Foundation, Wageningen, The Netherlands.

Lasso, C. 2002. Biological Diversity of Guianan Fresh and Brackish Water Fishes. *In*, Guayana Shield Conservation Priority Setting Workshop. April 5–8, Paramaribo, Suriname. Conservation International, Washington, DC.

Lim, B. and M. Engstrom. 2001. Species diversity of bats (Mammalia: Chiroptera) in Iwokrama Forest, Guyana, and the Guianan subregion: implications for conservation. Biodiversity and Conservation. 10: 613–657.

Lim, B.K. and Z. Norman. 2002. Rapid Assessment of Small Mammals in the Eastern Kanuku Mountains, Lower Kwitaro River Area, Guyana. *In*, Montambault, J.R. and

O. Missa (eds.). A Biodiversity Assessment of the Eastern Kanuku Mountains, Lower Kwitaro River, Guyana. RAP Bulletin of Biological Assessment 26. Conservation International, Washington, DC.

LOC (Library of Congress). 1992. Guyana Country Study. Library of Congress, Washington, DC.

Mayell, H. 2001. Biologists Document Rich Plant Life of Guyana to Aid Conservation. National Geographic News Online. Washington, DC.

Mittermeier, R.A., P. Robles-Gil, and C.G. Mittermeier. 1997. Megadiversity: Earth's Biologically Wealthiest Nations. Cemex, México.

Mittermeier, R.A., N. Myers, J.B. Thomsen, G.A.B. da Fonseca, and O. Silvio. 1998. Biodiversity Hotspots and Major Tropical Wilderness Areas: Approaches to Setting Conservation Priorities. Conservation Biology. 12(3): 516–520.

Mol, J. 2002. A Preliminary Assessment of the Fish Fauna and Water Quality of the Eastern Kanuku Mountains: Lower Kwitaro River and Rewa River at Corona Falls. In, Montambault, J.R. and O. Missa (eds.). 2002. A Biodiversity Assessment of the Eastern Kanuku Mountains, Lower Kwitaro River, Guyana. RAP Bulletin of Biological Assessment 26. Conservation International, Washington, DC.

NDS (National Development Strategy Secretariat). 1997. Environmental Policy. National Development Strategy, 3:18. Ministry of Finance, Georgetown, Guyana.

NewsWire. 2002. Joint venture nets $1.2 million for Vanessa's Marudi Gold project. News release: April 8. Canada NewsWire, Vancouver, Canada.

Parker, T.A., III, R.B. Foster, L.H. Emmons, P. Freed, A.B. Forsyth, B. Hoffman, and B.D. Gill (eds.). 1993. A biological assessment of the Kanuku Mountain region of southwestern Guyana. RAP Working Papers 5. Conservation International, Washington, DC.

Peres, C.A. and P.M. Dolman. 2000. Density compensation in neotropical primate communities: evidence from 56 hunted and nonhunted Amazonian forests of varying productivity. Oecologia 122:175–189.

Rettig, N.L. 1977. In quest of the snatcher. Audubon. 79(11): 26–49

Rettig, N.L. 1978. Breeding behavior of the Harpy Eagle (Harpia harpyja). Auk. 95(4): 629–643.

Rettig, N.L. 1995. Remote world of the Harpy Eagle. National Geographic. 187:40–49.

Reynolds, R., R. MacCulloch, M. Tamessar, and C. Watson. 2001. Preliminary checklist of the Herpetofauna of Guyana. The Biological Diversity of the Guianas Program, Smithsonian Institution, Washington, DC.

Sanderson, J. and L. Ignacio. 2002. Non-Volant Mammal Survey Results from the Eastern Kanuku Mountains, Lower Kwitaro River, Guyana. In, Montambault, J.R. and O. Missa (eds.). 2002. A Biodiversity Assessment of the Eastern Kanuku Mountains, Lower Kwitaro River,

Guyana. RAP Bulletin of Biological Assessment 26. Conservation International, Washington, DC.

Singh, J.G. 1994. The Enforcement Experience in Guyana on Exploitation of Natural Resources. In, Third International Conference on Environmental Enforcement. April 25–28, Oaxaca, México.

Stabroek News. 2001. Editorial: Fuzzy Planning. August 12. Georgetown, Guyana.

ter Steege, H. 1998. The use of large-scale forest inventories for a National Protected Area Strategy in Guyana. Biodiversity and Conservation. 7: 1457–1483.

ter Steege, H. 2001. National Vegetation Map of Guyana and explanatory notes. Guyana Forestry Commission and University of Utrecht, The Netherlands.

ter Steege, H. and G. Zondervan. 2000. A preliminary analysis of large-scale forest inventory data of the Guyana Shield. In, H. ter Steege (ed.), Plant Diversity in Guyana, with recommendations for a National Protected Area Strategy (Tropenbos Series 18). The Tropenbos Foundation, Wageningen, The Netherlands. pp. 35–54.

ter Steege, H., V.G. Jetten, A.M. Polak, and M.J.A. Werger. 1993. Tropical rain forest types and soil factors in a watershed area in Guyana. Journal of Vegetation Science. 4: 705–716.

ter Steege, H., M.J. Jansen-Jacobs, and V.K. Datadin. 2000. Can botanical collection assist in a national Protected Area Strategy in Guyana? Biodiversity and Conservation. 9: 215–240.

ter Welle, B.J.H., M.J. Jansen-Jacobs, A.R.A. Gorts-van Rijn, and R.C. Ek. 1987. Botanical exploration in the Northern Part of the Western Kanuku Mountains (Guyana). Inst. of Systematic Botany, Utrecht University, The Netherlands.

ter Welle, B.J.H., M.J. Jansen-Jacobs, A.R.A. Gorts-van Rijn, and R.C. Ek. 1990. Botanical exploration in the Northern Part of the Western Kanuku Mountains (Guyana). Annex: Identifications. Inst. of Systematic Botany, Utrecht University, The Netherlands.

ter Welle, B.J.H., M.J. Jansen-Jacobs, and P. P. Haripersaud. 2000. Botanical exploration in Guyana Corona Falls (Rewa River) Towards Essequibo River. University of Utrecht, Herbarium, The Netherlands.

Thomas, O. 1901. On a collection of mammals from the Kanuku Mountains, British Guiana. Annals and Magazine of Natural History. 7(7): 139–154.

Chapter 2

The Vegetation Along the Lower Kwitaro River on the Eastern Edge of the Kanuku Mountain Region, Guyana

Wilmer Díaz

ABSTRACT

During nine days of fieldwork the structure and composition of the diverse plant communities at two study sites, Pobawau Creek (at the confluence with the Kwitaro River) and Cacique Mountain, were inventoried and described. The mixed forests were dominated by species in the families Lecythidaceae, Chrysobalanaceae, Meliaceae, Sapotaceae, Fabaceae, Caesalpiniaceae, Apocynaceae, Euphorbiaceae, Annonaceae, and Violaceae. Peaks at Cacique Mountain support shrubby vegetation comprised of *Clusia* spp., *Erythroxylum* spp., terrestrial bromeliads, orchids, ferns, and aroids. The rocks, trunks, and branches of the shrubs are covered with mosses, epiphytic orchids, ferns, and aroids. Additional plant communities along the Lower Kwitaro River and Pobawau Creek are described in this report.

INTRODUCTION

The flora of the Guayana Shield is rich, with approximately 8,000 documented species; 50% of which are considered endemic to the region (Parker et al., 1993). The landscape of the Eastern Kanuku Mountains in southwestern Guyana has been little explored and supports large tracts of undisturbed forest in a diversity of habitats, with an impressive but understudied diversity of plants. Previous expeditions to the Kanuku Mountains and surrounding area, carried out by Marion Jansen-Jacob (Rewa River, Corona Falls, and Kanuku Mountains), Hans ter Steege, and Ben ter Welle from the Utrecht University (ter Welle et al., 1987, 1990, 2000; Gérard et al., 1996; Jansen-Jacobs and ter Steege, 2000; ter Steege, 2000) and Bruce Hoffman and David Clarke (Boggan et al., 1997; Hollowell et al., 2001; Hoke, 2002) from the Smithsonian Institution, have produced a list of 1,577 species, representing 26.3% of Guyana's flora. Additionally, the previous 1993 Rapid Assessment Program (RAP) expedition carried out in the Western Kanuku Mountains reported 251 plant species and emphasized the biological importance of this region (Parker et al., 1993). For this reason, Conservation International sponsored a second RAP expedition to the Eastern Kanuku Mountains to collect preliminary biological data, catalyze conservation efforts in Guyana, and create a framework for decision makers to further conservation efforts in southern Guyana.

MATERIALS AND METHODS

Transects were carried out to obtain a quantitative description of vegetation structure and community composition in mixed forests, following the methodology described by Foster et al. (1995). For each transect, 50 trees of three diameter categories were counted and identified (between 1 and 10 cm, between 10 and 30 cm, and greater than 30 cm), as well as 50 herbs. When possible, vouchers of the most common species were collected, as well as any others that were in flower. The specimens collected were placed in a plastic bag to be processed in the campsite as follows: the information relating to the specimen, habitat, and locality was recorded in a field book, then each specimen was numbered and placed in a sheet of newspaper with its respective number. When the stack reached approximately 30 cm high, the package was encased in newspaper, placed in a plastic bag and soaked with a 1:1 mixture of ethanol (95%) and water to preserve specimens for later processing in the herbarium. Vouchers were deposited at the University of Guyana Herbarium and the Herbario Regional de Guayana (GUYN) in the Jardín Botánico del Orinoco, Ciudad Bolívar, Venezuela. Final plant identification was completed through a collaboration between the Herbario GUYN, Herbario Universitario (PORT) at the Universidad Nacional Experimental de los Llanos Ezequiel Zamora, in Guanare, Portuguesa (Venezuela), with the assistance of specialists at the University of Wisconsin Herbarium (WIS) and the Marie Selby Botanical Gardens Herbarium in Sarasota, FL (SEL).

STUDY AREA

We conducted our study from 21–29 September 2001, representing the beginning of the dry season. Our sample sites included Pobawau Creek (3°16'3.1"N, 58°46'42.7"W) and Cacique Mountain (3°11'29.5"N, 58°48'42.0"W) 10 km southwest of Pobawau Creek. Both sites were located on the Kwitaro River in the Rewa River Basin, which is a tributary of the Rupununi River in the larger Essequibo drainage system and within southern Guyana's Region 9. Both sites were at approximately 120 m elevation (one peak at 450 m was surveyed). Vegetation was lowland seasonally inundated and *terra firma* evergreen tropical forest. River depth was high for the dry season, but dropped rapidly, falling approximately 1.5 m during our brief visit.

RESULTS

The distribution of vascular plants collected or observed in the RAP expedition study area, as well as the dominant elements of the plant communities are highlighted in Tables 2.1 and 2.2. Our preliminary results (based primarily on field work and pending final identification) indicate that the most important families (in term of numbers of species) are Fabaceae (13 spp.), Rubiaceae (12 spp.), and Meliaceae and Mimosaceae, both with 9 species. We documented 215 plant species, from 198 specimens and more observations, comprising 72 families (Appendix 2). Among these, 40 species appear to be new records for the Kanuku Mountains.

In general, the RAP team encountered mixed forests in flood plains and valleys that were physiologically and floristically similar to the tall evergreen non-flooded forests of the alluvial plains and interior hill-lands of the Venezuelan Guayana Shield, especially Delta Amacuro, in its extension from the Grande River to the Imataca Ridge and farther into the Amacuro-Cuyuní watershed (Berry et al., 1995). The forests found on the slopes of Cacique Mountain, however, resemble the tall evergreen premontane non-flooded forests of Cuyuní-Caroní lowlands, in the premontane zone that extends over the foothill landscape to the north of Lema Ridge and the area known as La Escalera. The shrubby vegetation on Cacique Mountain top is similar to that found in the granite outcrops of Cangrejo Mountain along the Caura River. The habitat encountered in the river meanders of the Lower Kwitaro can be compared to that found in the 2000 AquaRAP (freshwater biodiversity assessment) on the Lower Caura River, Venezuela (Rosales et al., *in press*). These comments should be considered preliminary, as our time in the field and for analyzing the data was brief. A more extensive comparison of the Eastern Kanuku flora to both the Western Kanuku Mountains and other parts of the Guayana Shield is recommended.

Table 2.1. The vascular plants of the RAP expedition study area are distributed in the following taxa:

Classes	Families	Genera	Species
Pteridophyta			
Filicopsida	8	12	13
Magnoliophyta			
Liliopsida	10	26	30
Magnoliopsida	54	105	172
Total	**72**	**143**	**215**

Table 2.2. Dominant elements of plant communities by family and genus in terms of number of species in order of abundance.

Families	# species	Genera	# species
Fabaceae	13	*Duguetia*	4
Rubiaceae	12	*Inga*	3
Meliaceae	9	*Piper*	3
Mimosaceae	9	*Matayba*	3
Apocynaceae	8	*Psychotria*	3
Annonaceae	7	*Pouteria*	3
Myrtaceae	6	*Rinorea*	3
Polypodiaceae	6	*Geonoma*	2
Caesalpiniaceae	5	*Scleria*	2
Poaceae	5	*Micrograma*	2

POBAWAU CREEK SITE: CONFLUENCE BETWEEN KWITARO RIVER AND POBAWAU CREEK

Overall the forest was in good condition. There was no evidence of logging or clearings. The only sign of human disturbance was that balata trees (*Manilkara* sp.) showed cuts from latex bleeding on their trunks.

Mixed Forest

This type of forest was present on flat generally *terra firma* terrain along the Kwitaro River, occasionally flooded in the wet season. The forest had an average height of 30–35 m, and was floristically rich in species as compared to Mora forests in other seasonally flooded parts of Guyana. In a transect of about 280 m, we found approximately 27 species of trees greater than 30 cm diameter, the most abundant of which were *Bertholletia excelsa* (14%), *Eschweilera* sp. (6%), *Catostemma fragans* (6%), and an unidentified species of Sapotaceae (6%). In the category of trees between 10 and 30 cm diameter, the most abundant of the 24 species were *Sagotia racemosa* (18%), *Eschweilera* sp. (10%), *Carapa guianensis* (10%), *Hirtella* sp. (8%), and *Eschweilera* sp. 2 (8%). The shrubby layer, made up of individuals between 1 and 10 cm diameter, was dominated by *Rinorea* sp. (28%), *Duguetia* sp. (20%), *Eschweilera* sp. (8%), *Hirtella* sp. (8%), and *Sterculia* sp. (8%), among the 15 species counted. In the herbaceous layer the most common species were *Geonoma* sp. (32%), *Adiantum* sp. (20%), *Selaginella* sp. (12%), and *Ischnosiphon* sp. (10%). The abundance of *Bertholletia excelsa* in this forest may be associated with past human activity (ter Steege, 1998).

Tree fall gaps in the forest contained many lianas, principally *Smilax* sp., *Doliocarpus* sp., and *Bauhinia* sp., among others. There were few tree palms. Several epiphytes were observed on the branches of canopy trees, primarily *Vittaria* sp., *Pleopeltis* cf. *percussa*, *Huperzia* cf. *dichotoma*, *Microgramma* sp., *Philodendron* sp., and two species of Orchidaceae. The hemi-epiphyte *Heteropsis* sp., highly valued as bush-rope and for use in basket weaving, was very common on tree trunks.

Another transect of about 330 m was made beside the Kwitaro River. The average height of this forest was 30–35 m. The results showed 19 species of more than 30 cm diameter, the most abundant of which were *Eschweilera* sp. 1 (24%), *Inga* sp. (12%), *Eschweilera* sp. 2 (12%), and *Catostemma fragans* (10%). In the sub-canopy (trees between 10 and 30 cm diameter), the most abundant species were *Sagotia racemosa* (22%), *Inga* sp. (20%), *Licania* sp. (6%), *Eschweilera* sp. (6%), *Catostemma fragans* (6%), and *Eschweilera* sp. 2 (6%). The shrubby layer was highly dominated by *Rinorea* sp. (44%) and *Sagotia racemosa* (20%). The most abundant grass was *Adiantum* sp. (88%). This forest also presented tree fall gaps, where the species *Clathrotropis macrocarpa* and *Inga* sp. were the most common, along with vines of the families Bignoniaceae, Fabaceae, Mimosaceae, Sapindaceae, a different species of *Smilax*, *Dolliocarpus* sp., and *Bauhinia* sp. Compared with the forest from the first transect, the primary difference was the absence of *Bertholletia excelsa* and *Geonoma* sp.

River Meander Successional Forest

On the sandy bar of the Kwitaro near the mouth of Pobawau Creek, the predominant species was *Inga* cf. *meissneriana*, with its trunk growing horizontally toward the river and forming a barrier of 5 to 10 m wide. Behind this barrier, the vegetation consisted of herbs and small shrubs 20–150 cm high of *Cyperus* sp., *Leptochloa* sp., *Diodia* sp., *Solanum* sp., and *Ichnanthus* sp. In the depression, or backwater area behind the riverbank, the most common species were *Calathea* sp., *Costus* sp., *Tabernaemontana* sp., *Bytneria* sp., intertwined with vines of *Cissus* sp., *Dalechampia* cf. *affinis*, *Passiflora* sp., *Entada* sp., and others in the Malpighiaceae and Fabaceae families. Among the shrubs it was common to see *Bixa orellana*, *Piper* sp., *Annona* sp., *Casearia* sp., and a bamboo (possibly *Guadua*). The most frequent and abundant trees were *Triplaris* cf. *weigeltiana* and *Cecropia peltata*, and some scattered specimens of *Cordia* sp. and *Iryanthera* sp.

River Banks

On the stable, lower mud flats of the Pobawau Creek the characteristic tree was *Psidium* sp., but it was also common to see *Genipa spruceana*, *Alibertia* sp., *Macrolobium* sp., *Homalium* sp., *Tournefortia* sp., and *Montrichardia arborescens*. On the edges of the forest there were some characteristic tree species such as *Croton* sp., *Matayba* sp., *Talisia* sp., *Mabea* sp., and *Maytenus* sp. It was also common to see vines like *Cuervea* sp., *Combretum* sp., and others in the Fabaceae and Bignoniaceae families.

CACIQUE MOUNTAIN SITE: LOWER KWITARO RIVER AND MOUNTAIN REGION

Cacique Mountain Top

One mountaintop was surveyed which had a granite outcrop about 400–450 m high and a flattened peak of approximately 5 by 10 m. In this small area, the vegetation was distinct from the surrounding forest, composed of shrubs 5–8 m tall such as *Clusia* spp., *Erythroxylum* spp., *Jacaranda copaia*, *Terminalia* sp., and one Myrtaceae. Mosses and epiphytes covered the trunk and branches of these shrubs, as well as the rocks. Between the species observed, there were six different Orchidaceae, two Araceae, and a single species each of *Pecluma* sp., *Polypodium* sp., and *Epiphyllum* sp. The herbaceous layer was almost completely composed of *Pitcairnia* sp. and *Bromelia* sp., but there were also three species of Orchidaceae, two Araceae, one Rubiaceae, and two Cyperaceae. There were some vines of the family Malpighiaceae and a vine-like cactus of genus *Hylocereus*.

Cacique Mountain Slope

The hill slopes were a Mixed Forest with an average height of 20–25 m, reflecting the species richness of this forest. Twenty tree species greater than 30 cm diameter were found in an approximately 250 m long transect. In this category the most common species were *Carapa guianensis* (14%), *Catostemma fragans* (12%), *Licania* sp. (12%), *Pouteria* spp. (12%), and *Trattinickia* sp. (6%). In the sub-canopy (10–30 cm diameter), the same species found in the canopy were also the most abundant: *Catostemma fragans* (12%), *Carapa guianensis* (10%), and *Licania* sp. (10%), with the addition of *Pouteria* spp. (8%) and *Eschweilera* sp. (6%). The shrub layer (1–10 cm diameter) was dominated by *Duguetia* spp. (30%), *Licania* sp. (16%), and a Fabaceae sp. (10%). Among the most common herbs, *Selaginella* sp. (42%), *Olyra* sp. (26%), and *Ischnosiphon* sp. (16%) formed the major part of this layer, but it was not unusual to see *Phenakospermun guyannense* (6%) and *Geonoma* sp. (6%). No tree fall gaps were found. Lianas were infrequent, and of those seen, the diameter of the stem was more than 30 cm. Epiphytes were also scarce.

Another transect was carried out to characterize the forest of a slope close to the floodplain. The forest in this case was 20% *Eschweilera* spp., 18% *Pouteria* spp., 8% *Licania* sp., and 8% *Trattinickia* sp. in the canopy. In the sub canopy, the most common species were *Clathrotropis* cf. *macrocarpa* (14%), *Eschweilera* sp. (12%), *Aspidosperma* sp. (10%), *Licania* sp. (8%), and *Catostemma fragans* (6%). The shrubby layer was dominated by *Rinorea* spp. (20%), *Duguetia* spp. (12%), *Guarea* sp. (10%), Lauraceae sp. (8%), and *Sterculia* sp. (8%), and the most abundant herbs were *Ischnosiphon* sp. (66%) and *Selaginella* (24%). Both palms and the epiphytes were very scarce. No tree fall gaps were observed, but lianas were present.

Valleys

The forest in the valley differs significantly from that of the slope. In this case, we found a strong presence of palm trees not seen in other forests (*Euterpe* sp., *Oenocarpus* sp, *Astrocaryum* sp.), probably because of the creeks. This forest was dominated by a single species, especially noteworthy because the other forests were highly mixed. A transect of about 250 m long showed that *Clathrotropis* cf. *macrocarpa* accounted for 40% of the trees greater than 30 cm diameter and 38% of the trees between 10 and 30 cm diameter. Other common canopy species included *Carapa guianensis* (14%) and *Pouteria* spp. (10%). In the sub canopy, *Sagotia racemosa* (10%), *Carapa guianensis* (8%), *Clathrotropis* cf. *brachypetala* (8%), and *Eschweilera* sp. (6%) were the most frequently observed. However, among the individuals between 1 and 10 cm diameter, the most common species were *Sagotia racemosa* (20%), *Duguetia* spp. (16%), *Clathrotropis* cf. *macrocarpa* (8%), and *Licania* sp. (6%). The most common herbs were *Selaginella* sp. (27%), *Geonoma* sp. (16%), and *Ischnosiphon* sp. (10%). Tree fall gaps were observed, filled with vines of the families Mimosaceae, Passifloraceae (*Passiflora* sp.),

and Caesalpiniaceae (*Bauhinia* sp.) as well as the herb *Olyra* sp. The epiphytes were not common, but the hemi-epiphytic *Heteropsis* sp. was always present, along with another aroid, *Anthurium* sp.

Flood Plains

A transect was made in the flood plain at the beginning of the slope of Cacique Mountain, in a forest with an average height of 25–30 m. The results show that the most common species with a diameter greater than 30 cm were *Clathrotropis* cf. *macrocarpa* (14%), a Caesalpiniaceae (possibly *Bocoa*) (12%), *Eschweilera* sp. (10%), *Pterocarpus* cf. *rohrii* (10%), and *Eschweilera* sp. 2 (10%). In the sub-canopy (10–30 cm diameter), the most common species were *Sagotia racemosa* (22%), Meliaceae sp. (14%), *Bocoa?* sp. (12%), and *Clathrotropis* cf. *macrocarpa* (8%). The shrub layer consisted of 22% *Sagotia racemosa*, 18% *Rinorea* spp., 12% *Guarea* sp., 10% *Duguetia* spp., and 10% *Bocoa?* sp. The most common herbs were *Geonoma* sp. (48%) and *Selaginella* sp. (26%). Palms were present, as were tree fall gaps covered with vines and epiphytes (common on higher branches of canopy trees).

Another transect was made in the 25–30 m forest beside the Kwitaro River. In the canopy (diameter greater than 30 cm) the most common species were *Bocoa?* sp. (20%), *Catostemma fragans* (16%), Lauraceae sp. (10%), *Eschweilera* sp. (6%), and Meliaceae sp. (6%). The sub-canopy was dominated by the same species of *Bocoa?* sp. (22%), *Eschweilera* sp. (10%), and *Pouteria* spp. (10%). In the shrub layer, the most common species were *Rinorea* spp. (18%), *Duguetia* spp. (16%), *Bocoa?* sp. (10%), and *Guarea* sp. (10%). The herbaceous layer showed predominately *Ischnosiphon* sp. (58%), *Adiantum* sp. (18%), and *Geonoma* sp. (10%). Tree palms were scarce (some *Euterpe* sp. were recorded), as were the epiphytes, with the exception of *Anthurium* sp. and the omnipresent *Heteropsis* sp. Tree fall gaps were common in this area.

CONCLUSIONS

The vegetation in the study area was composed mainly of mixed forests on slopes, in valleys and flood plains. The most common species were *Licania* sp., *Carapa guianensis*, *Eschweilera* spp., *Pouteria* spp., *Catostemma fragans*, *Aspidosperma* sp., *Trattinickia* sp., *Sagotia racemosa*, *Duguetia* spp., and *Rinorea* spp. In the river meanders, the most common tree species were *Triplaris* cf. *weigeltiana*, *Cecropia peltata*, and *Cordia* sp. On the stable, lower mud flats of the Pobawau Creek the characteristic species were *Psidium* sp., *Genipa spruceana*, *Alibertia latifolia*, *Macrolobium acaciifolium*, *Homalium racemosum*, *Tournefortia* sp., and *Montrichardia arborescens*.

The abundance of *Bertholletia excelsa* in one of the forests sampled may be associated with past human activity (ter Steege, 1998). No logging or clearings were observed. The only visible trace of current human disturbance was in the form of balata tree bark (*Manilkara* sp.) cut to collect latex.

On the Cacique Mountain, the vegetation on the summit was completely different from the surrounding forest. The forest in the valley differs significantly from that of the slope, probably resulting from the presence of creeks. Palm trees, not seen in other forests (*Euterpe* sp., *Oenocarpus* sp, *Astrocaryum* sp.), were abundant in this habitat. The most common species in the herbaceous layer of the forest were *Selaginella* sp., *Ischnosiphon* spp., *Geonoma* sp., and *Adiantum* sp.

RECOMMENDATIONS

More time is needed to visit and describe the plant communities on top of mountains, slopes, and intermountain valleys to have a better idea of their floristic richness, habitats, and endemism—especially on top of mountain peaks over 500 m high. Additional studies to compare the flora of the Eastern Kanukus to the Western Kanukus are also recommended.

The study area possesses a great deal of undisturbed forest, which appears not to be threatened by mining or logging at present. The scenic beauty and wilderness of the area presents an excellent opportunity for protection. Studies of the area indicate that the diverse ecosystems of the Kanuku Mountains support a large percentage of the biodiversity of the country. The Takutu bridge at Lethem, already under construction, will open this now remote region to Brazil, therefore it is critical to place the Kanuku Mountain area under some status of legal protection.

LITERATURE CITED

Berry P.E., B.K. Holst, and K. Yatskievych (eds.). 1995. Flora of the Venezuelan Guyana, Vol. 1. Introduction. Missouri Botanical Garden and Timber Press, Saint Louis, MO.

Boggan, J., V. Funk, C. Kelloff, M. Hoff, G. Cremers, and C. Feuillet. 1997. Checklist of the Plants of the Guianas, 2nd Edition. Biological Diversity of the Guianas Program, Smithsonian Institution, Washington, DC. 238 pp.

Foster, R.B., N.C. Hernández E., E.K. Kakudidi, and R.J. Bunham. 1995. A variable transect method for rapid assessment of tropical plant communities. The Field Museum, Unpublished Working Paper. Chicago, IL.

Gérard, J., R.B. Miller, and B.J.H. ter Welle. 1996. Major Timber Trees of Guyana. Timber Characteristics and Utilization. Tropenbos Series 15. National Herbarium Nederland, Utrecht University Branch, The Netherlands.

Hoke, P. (ed.). 2002. Combined Species Plant List for Southern Guyana: Utrecht University, Smithsonian, and Rapid Assessment Program data sources. Unpublished Database. Conservation International, Washington, DC.

Hollowell, T., P. Berry, V. Funk, and C. Kelloff. 2001. Preliminary Checklist of the Plants of the Guyana Shield (Venezuela: Amazonas, Bolívar, Delta Amacuro; Guyana; Surinam; French Guiana). Volume 1: Acanthaceae–Lythraceae. Flora of the Venezuelan Guayana series. Smithsonian Institution, Washington, DC.

Jansen-Jacobs, M. and H. ter Steege. 2000. Southwest Guyana: a complex mosaic of savannahs and forests. *In*, H. ter Steege (ed.). Plant Diversity in Guyana, with recommendations for a National Protected Area Strategy (Tropenbos Series 18). pp. 147–157. The Tropenbos Foundation, Wageningen, The Netherlands.

Parker, T.A., III, R.B. Foster, L.H. Emmons, P. Freed, A.B. Forsyth, B. Hoffman, and B.D. Gill (eds.). 1993. A biological assessment of the Kanuku Mountain region of southwestern Guyana. RAP Working Papers 5. Conservation International, Washington, DC.

Rosales, J., M. Bevilacqua, W. Díaz, R. Pérez, D. Rivas, and S. Caura (*in press*). Riparian Vegetation Communities of the Caura River. *In*, B. Chernoff, K. Riseng, A. Machado-Allison, and J.R. Montambault (eds.). A Biological Assessment of the Aquatic Ecosystems of the Río Caura Basin, Bolívar State, Venezuela. RAP Bulletin of Biological Assessment, 28. Conservation International, Washington, DC.

ter Steege, H. 1998. The use of large-scale forest inventories for a National Protected Area Strategy in Guyana. Biodiversity and Conservation. 7: 1457–1483.

ter Steege, H. (ed.). 2000. Plant Diversity in Guyana, with recommendations for a National Protected Area Strategy (Tropenbos Series 18). pp. 35–54. The Tropenbos Foundation, Wageningen, The Netherlands.

ter Welle, B.J.H., M.J. Jansen-Jacobs, A.R.A. Gorts-van Rijn, and R.C. Ek. 1987. Botanical exploration in the Northern Part of the Western Kanuku Mountains (Guyana). Inst. of Systematic Botany, Utrecht University, The Netherlands.

ter Welle, B.J.H., M.J. Jansen-Jacobs, A.R.A. Gorts-van Rijn, and R.C. Ek. 1990. Botanical exploration in the Northern Part of the Western Kanuku Mountains (Guyana). Annex: Identifications. Inst. of Systematic Botany, Utrecht University, The Netherlands.

ter Welle, B.J.H., M.J. Jansen-Jacobs, and P.P. Haripersaud. 2000. Botanical exploration in Guyana Corona Falls (Rewa River) Towards Essequibo River. University of Utrecht, Herbarium, The Netherlands.

Chapter 3

A Preliminary Assessment of the Fish
Fauna and Water Quality of the Eastern
Kanuku Mountains: Lower Kwitaro River
and Rewa River at Corona Falls

Jan H. Mol

ABSTRACT

The water quality and fish fauna of the Eastern Kanuku Mountains in Guyana, South America, were studied during a 10-day Rapid Assessment Program (RAP) expedition organized by Conservation International. The water quality of the Rewa and Kwitaro rivers that drain the Eastern Kanuku Mountains exhibited the characteristics of an Amazonian clear-water stream. The Rewa River had a slightly lower conductivity and a higher transparency than the Kwitaro River. Small tributary rainforest creeks in the study area were more acidic than the large rivers. Although sampling conditions were not optimal, we collected 113 fish species in the catchments of the Lower Kwitaro River and the Upper Rewa River at the Corona Falls rapid. Among the 45 large-sized (>15 cm Standard Length) food fishes, the arapaima (*Arapaima gigas*) and the arawana (*Osteoglossum bicirrhosum*) are perhaps most notable. Many ubiquitous small-sized fishes (characins, knife-fishes, catfishes, and cichlids) from the leaf litter and woody debris habitats were absent from the catches as a result of sub-optimal sampling conditions (i.e., the high water level and possibly the bright moonlight). The high-altitude rocky headwater streams habitat was not encountered during the expedition. A second expedition in the low-water season is strongly recommended in order to arrive at a more complete and representative listing of the fish fauna.

INTRODUCTION

Conservation International has been involved in setting conservation priorities for areas of high biological diversity in the Guayana Shield (e.g., Parker et al., 1993). The first Rapid Assessment Program (RAP) survey in the Kanuku Mountain region of southwestern Guyana (Parker et al., 1993) suggested that the Eastern Kanuku Mountains might have special conservation value because the area is very remote and is suspected to have an even higher vertebrate biodiversity than the Western Kanukus. The Guayana Shield is already known to be one of the most species-rich freshwater ecosystems in the world with a high degree of fish species endemism (Gery, 1969) and its streams still largely in a pristine state. Freshwater ecosystems were, however, not included in the 1993 RAP expedition in Guyana.

The freshwater ecosystems of the Kanuku Mountains in the Rupununi area have not been studied in depth, if at all (Eigenmann, 1912; but see Lowe-McConnell, 1964, 2000 on the ecology of the fishes from the Rupununi Savannah). The Kanuku Mountains are situated in the catchments of the Rupununi and Rewa rivers (with its main tributary the Kwitaro River). The Rupununi River itself is a tributary of the larger Essequibo River basin. The Essequibo, Rewa, and Kwitaro rivers may be classified as clear-water rivers (terminology of Sioli, 1950), but the Rupununi River shows some characteristics of a white water river (ANSP, 1999). Large areas of the Rupununi savannah are flooded each year in the rainy (high-water) season, and in certain years the savannah connects the Amazon River (Branco River) and the Essequibo River basins (Rupununi River). This explains the presence of some large Amazonian fishes and rep-

A Preliminary Assessment of the Fish Fauna and
Water Quality of the Eastern Kanuku Mountains:
Lower Kwitaro River and Rewa River at Corona Falls

tiles in the Rupununi and Kanuku Mountain regions (e.g., *Arapaima gigas, Osteoglossum bicirrhosum, Colossoma bidens, Melanosuchus niger*). Due to this mixing effect, a high freshwater biodiversity may be expected in the Kanuku Mountain area as compared to other areas of the Guayana Shield (e.g., in Suriname and French Guiana).

In the period 20–30 September 2001, Conservation International carried out a second Rapid Assessment Program expedition in the Eastern Kanuku Mountains, specifically the Lower Kwitaro River and the Upper Rewa River at Corona Falls. Although identification of the specimens we collected is complete, these results are considered preliminary due to limited sampling effort and water level conditions that were suboptimal for a full survey of the fish fauna (e.g., compare our list of approximately 100 species with the fish fauna list of some 400 species from the Iwokrama rainforest in southern Guyana (ANSP, 1999) after several sampling expeditions at low water. We expect that at least 300 species are actually present in the area where we collected). For these reasons, I hereby present the preliminary results of the water quality and fish fauna survey of the Eastern Kanuku Mountains.

MATERIALS AND METHODS

Water quality and fish fauna were assessed at eight sampling sites:

(S1) Pobawau Creek (3º16'3.1"N, 58º46'42.7"W)
(S2) a bay in the Kwitaro River, 500 m upstream of Pobawau Creek
(S3) a shallow, short cut connection between two bends of the Kwitaro River
(S4) the Kwitaro River at Cacique Mountain camp (3º11'29.5"N, 58º48'42.0"W)

(S5) a lake connected to the Kwitaro River (approximately 700 m upstream of S4)
(S6) a small, unnamed forest creek (elevation 110 m) at the base of the Cacique Mountain slope
(S7) the Rewa River at the Corona Falls rapid (3º11'35.0"N, 58º48'39.6"W)
(S8) an unnamed forest creek at the Corona Falls rapid (same coordinates as S7).

Water quality was assessed with an Oakton pH/Conductivity/Degrees Celsius meter and a Secchi disk. Fish were collected both during the day and night using several methods: gill nets (1.25 and 1.75 cm unstretched mesh size), baited minnow traps, angling, hook and line, a 3 m-long seine (2 mm mesh size) and dip net (Table 3.1). The high water level and the bright moonlight affected the catches negatively, especially catches of small-sized fishes (see Discussion section). Large fish species (>15 cm Standard Length; food fishes) were identified in the field using keys in Gery (1977), Le Bail et al. (2000), and Keith et al. (2000); the small-sized species were identified at a later stage in the laboratory of the University of Suriname. Additional information on the occurrence of large food fishes in the study area was obtained from three informants: Duane de Freitas (Dadanawa Ranch, southern Rupununi savannah), Ashley Holland, and Guy Rodrigues (Shea village); the independent information from the informants was checked against each other and against personal observations.

RESULTS

The pH, conductivity, and visibility of the Kwitaro River, the Rewa River, and forest creeks in the Eastern Kanuku Mountain region (Table 3.2) was within the range of observations known for Amazonian streams (Furch, 1984): the pH varied between 5.60 and 6.42, the conductivity was

Table 3.1. Fishing methods and sampling effort per collection site.

Site	Gill net (net*night/net*day)	Seine (hours)	Minnow trap (baited) (trap*night/trap*day)	hook & line / angling (hours)	dip net (night) (hours)
S1	9 / 9	4	3 / 3	2	4
S2	5 / 5	-	- / -	3	-
S3	- / -	3	- / -	-	-
S4	2 / 2	-	- / -	8	1
S5	16 / 14	- ,	- / -	2	2
S6	- / -	3	- / -	-	-
S7	- / -	-	- / -	6	1
S8	1 / 1	1	- / -	-	2

Table 3.2. Water quality data of sampling sites in the Kwitaro and Rewa River systems.

Site	Date	Time	Stream width (m)	Water depth (m)	pH	Conductivity (µS cm⁻¹)	Water temperature (°C)	Secchi visibility (cm)
S1	21/09	08:30	5	>3	5.96	22.6	24.4	95
S1a*	23/09	07:00	-	1.5	5.72	21.8	24.8	109
S2	21/09	13:30	-	-	6.42	31.5	29.5	32
S3	21/09	14:00	5	2	6.39	31.1	30.0	21
S4	22/09	09:15	30	-	6.34	33.8	25.4	34
S5	24/09	16:00	-	4	5.76	32.4	29.4	57
S6	25/09	10:00	1	25	5.60	32.9	25.1	>25
S6a**	25/09	11:00	2.5	50	5.88	34.2	25.3	>50
S7	26/09	11:30	50	-	6.35	18.3	28.0	74
S8	26/09	14:00	3.5	3	5.87	24.3	28.8	202

* Pobawau Creek flooded forest
** second tributary of same forest creek at the foot of Cacique Mountain

low (18.3–34.2 µS cm⁻¹), and the transparency (Secchi disk visibility) varied between 21 and 202 cm. Water temperature ranged between 24.4° C in the Pobawau Creek early in the morning and 30.0° C in a shallow short-cut connection of the Kwitaro River (S3; 14.00 h). The pH of the forest creeks was slightly lower than the pH of the rivers (Table 3.2). The conductivity of the Rewa River was lower than the conductivity of the Kwitaro River, whereas the Secchi disk visibility was higher in the Rewa River than in the Kwitaro River.

We collected 113 fish species from the streams draining the Eastern Kanuku Mountains (Appendix 3); 45 species may be considered food fishes (>15 cm Standard Length). The arapaima (*Arapaima gigas*), arawana (*Osteoglossum bicirrhosum*), and red paku (*Colossoma bidens*) were not collected or observed in the area, but their presence was confirmed by at least two of the three informants.

DISCUSSION

The water quality of the Rewa and Kwitaro rivers did not differ from the water quality of Amazonian clear-water rivers (Furch, 1984). Rivers draining the Guayana Shield are characteristically either clear-water rivers (low conductivity, almost neutral pH, and high transparency) or black-water rivers (e.g., Rio Negro; coffee colored, acidic, high concentrations of humid substances, low conductivity, and low oxygen content). The black-water rivers often originate in white-sand savannahs, which were not observed in the Eastern Kanuku Mountains. The Amazonian rainforest creeks are often slightly acidic with an extremely low conductivity

(Furch, 1984); this agrees reasonably well with our results of the water quality of the Kanuku Mountain forest creeks.

The water level in the Rewa and Kwitaro Rivers and in Pobawau Creek was still very high during the sampling period. Although the high water level of September 2001 was not normal for the time of the year, it was also not unusual (Duane de Freitas, pers. comm.). The high water level affected the sampling of fishes negatively in two ways. First, forests adjoining the Pobawau Creek and the Rewa and Kwitaro rivers were still inundated and the fishes were widely dispersed in the flooded forest (and therefore not as easily collected as in the low-water season when they are "locked up" in lakes and in pools in the stream channel). Second, the preferred sampling method to collect small-sized species, the small-meshed drag seine, could not be used efficiently because the water in the creeks was often too deep to walk through the creek and drag the seine. The bright moonlight also affected fish sampling negatively because in the moonlight the fishes can see and avoid the nets at night and/or their movements are more restricted (they remain essentially in one place and so do not encounter the set gill nets). Although the fish fauna of the Eastern Kanuku Mountains was sampled under sub-optimal conditions (high water level and full moon) and during a short period (10 days) we still collected/observed 113 fish species.

A more extensive collection effort in the Iwokrama Rainforest yielded 408 fish species (ANDP, 1999); with 345 species in the Essequibo River drainage. Many fish species of the Negro River in Brazil (Goulding et al., 1988) are also found in the Essequibo River (ANDP, 1999) due to the connections with the seasonally flooded Rupununi Savannah (Lowe-McConnell, 1964). The Rewa, Kwitaro, and Rupu-

A Preliminary Assessment of the Fish Fauna and
Water Quality of the Eastern Kanuku Mountains:
Lower Kwitaro River and Rewa River at Corona Falls

nuni Rivers are all part of the Essequibo River system; thus, it is likely that additional sampling efforts will considerably increase the number of fish species recorded for the Eastern Kanuku Mountains.

A listing of the very diverse fish fauna of a Guayana Shield watershed based on a relatively short sampling period of 10 days is obviously not complete, but the high water level (and to a lesser extent the bright moonlight) has probably resulted in a bias in the catches toward the large-sized species. This is unfortunate because most of the fish biodiversity in the Neotropics is related to the small-sized catfishes and characoids (Weitzman and Vari, 1988; Goulding et al., 1988). In addition, the important high-altitude rocky headwaters habitat, with potentially endemic species, was not surveyed during this expedition. An additional inventory of fishes in the Kanuku Mountains later in the dry season (late January or February) and under new moon conditions would display a more representative picture of fish fauna of the Eastern Kanuku Mountains and is strongly recommended.

The sampling effort during this RAP expedition was rather restricted and it is impossible to comment on to what extent the freshwater fish fauna of Guyana is present in the Eastern Kanuku Mountains and/or whether some fishes of the Kanuku Mountains are endemic to Guyana or even new to science. With respect to the large-sized fishes, I did not find a difference between the fish fauna of the Eastern Kanuku Mountains and the fish fauna of the Rupununi Savannah (Lowe McConnell, 1964), but differences among the two fish faunas are expected with respect to the small-sized species.

Although fishing and hunting is reported in the area in the dry season (Duane de Freitas, pers. comm.), the aquatic habitats did not seem to be disturbed to any large extent (this may change in the future, see below). The status of the fish populations in the area (e.g., *Arapaima gigas*) remains to be evaluated. However, over-fishing of *Arapaima gigas* for export to Brazil is reported (Duane de Freitas, pers. comm.). Mr. Ashley Holland (pers. comm.) held that the large food fishes are more abundant in the Rupununi River than in the Rewa/Kwitaro Rivers. This may be related to differences in the habitats of the two river systems (more lakes, river bays, and extensive flooding of savannahs along the Rupununi River) and/or a higher nutrient content of the Rupununi River associated with the regular burning of the adjoining savannahs (filamentous algae were observed in many locations in the Rupununi and these were apparently absent in the Rewa/Kwitaro). Next to over-fishing of the arapaima, the main threats to the fish fauna appear to be associated with small-scale gold mining (illegal Brazilian *garimpeiros*) and large-scale logging by Asian companies. Both activities can easily result in siltation of streams affecting a shift in the fish faunas (Mol and Ouboter, submitted) and smothering aquatic Podostemaceae plants in the rapids (with negative impacts on plant-eating characoids and young fishes that seek shelter in the plants).

RECOMMENDATIONS

- Conduct a second fish-sampling expedition in the low-water season (January–February) to arrive at a more complete and representative list of the fish biodiversity of the Eastern Kanuku Mountains, especially with respect to small-sized species (Appendix 3).

- Initiate studies on the biology, ecology, and culture of the world's largest freshwater fish, *Arapaima gigas*, a flagship species of the Kanuku Mountain area (e.g., investigate possibility of aging the arapaima by reading year marks in otoliths, thus enabling the construction of growth curves, the estimation of minimum age at maturity and maximum age, and the relation between differences in growth rate and environmental variables such as different habitats and climatic conditions).

- Continue and expand outreach programs to support existing legislation to protect the endangered arapaima and promote its culture; an aquaculture station with arapaima may also attract ecotourists.

- Promote the almost pristine Kanuku Mountains ecosystems to potential ecotourists in the United States and Europe (e.g., produce videos or leaflets with color photographs of the forest, savannah, rapids, and creek landscapes, the Amerindian communities and way of life, the biology and fisheries flagship species).

ACKNOWLEDGMENTS

I thank the informants Duane de Freitas, Ashley Holland, and Guy Rodrigues for sharing their extensive knowledge of the Kanuku fish fauna. Justin de Freitas and Asaph Wilson assisted with the sampling of fish. The smooth organization of the expedition to the Eastern Kanuku Mountains was due to Jensen Montambault, Olivier Missa, and Eustace Alexander.

LITERATURE CITED

ANSP (Academy of Natural Sciences Philadelphia). 1999. The Vertebrate Fauna of the Iwokrama Forest. Fishes. Final report from work carried out in the Iwokrama Forest by the Academy of Natural Sciences of Philadelphia 1996–1998. Academy of Natural Sciences Philadelphia, Philadelphia. Pp. 69–93.

Eigenmann, C.H. 1912. The freshwater-fishes of British Guiana. Mem. Carnegie Mus. 5: 1–578.

Furch, K. 1984. Water chemistry of the Amazon basin: the distribution of chemical elements among freshwaters. *In*, H. Sioli (ed.). The Amazon: limnology and landscape ecology of a mighty tropical river. Junk, Dordrecht, The Netherlands. pp. 167–199.

Gery, J. 1969. The freshwater fishes of South America. *In*, E.J. Fittkau, et al. (eds.). Biogeography and ecology in South America. Junk, The Hague, The Netherlands. Pp 828–848.

Gery, J. 1977. Characoids of the world. T.F.H., Neptune City.

Goulding, M., M.L. Carvalho, and E.G. Ferreira. 1988. Rio Negro: rich life in poor water. SPB, The Hague, The Netherlands.

Keith. P., P.Y. Le Bail, and P. Planquette. 2000. Atlas des poissons d'eau douce de Guyane. Tome 2, fascicule I. Batrachoidiformes, Mugiliformes, Beloniformes, Cyprinodontiformes, Synbranchiformes, Perciformes, Pleuronectiformes, Tetraodontiformes. Museum d'Histoire Naturelle, Paris.

Le Bail, P.Y., P. Keith, and P. Planquette. 2000. Atlas des poisons d'eau douce de Guyane. Tome 2, fascicule II. Siluriformes. Museum d'Histoire Naturelle, Paris.

Lowe-McConnell, R.H. 1964. The fishes of the Rupununi savanna district of British Guiana, South America. Journal of the Linnaean Society (Zoology) 45: 103–144.

Lowe-McConnell, R.H. 2000. Land of waters: explorations in the natural history of Guyana, South America. The Book Guild, Sussex, United Kingdom.

Mol, J.H. and P.E. Ouboter (submitted). Downstream effects of erosion from small-scale gold mining on the stream habitat and fish community of a neotropical rainforest creek. Submitted to Conservation Biology (13 February 2002).

Parker, T.A., III, R.B. Foster, L.H. Emmons, P. Freed, A.B. Forsyth, B. Hoffman, and B.D. Gill (eds.). 1993. A biological assessment of the Kanuku Mountain region of southwestern Guyana. RAP Working Papers 5. Conservation International, Washington, DC.

Sioli, H. 1950. Das Wasser im Amazonasgebiet. Forschung und Fortschritt 26: 274–280.

Weitzman, S.H. and R.P. Vari. 1988. Miniaturization in South American freshwater fishes: an overview and discussion. Proceedings of the Biological Society of Washington 101: 444–465.

Chapter 4

Avifauna of the Eastern Edge of the Eastern Kanuku Mountains, Lower Kwitaro River, Guyana

Davis W. Finch, Wiltshire Hinds, Jim Sanderson, and Olivier Missa

ABSTRACT

The RAP expedition to the Eastern Kanuku Mountains on the Lower Kwitaro River recorded 175 bird species. Here we combine these results with the recent fieldwork of Davis Finch in the same area. Our results show that the known bird fauna of this stretch of the Lower Kwitaro River now numbers 264 species, including 39 species considered uncommon in the Neotropics and two rare species, namely the Harpy Eagle (*Harpia harpyja*) and the Orange-breasted Falcon (*Falco deiroleucus*). The Kanuku Mountains, including both the western and eastern ranges, support an impressively diverse avifauna, with 419 species recorded, or about half the total presently known for the country. Because of its pristine environmental conditions and the presence of many bird species endemic to the Guayana Shield, the conservation of this area should be considered a high priority.

INTRODUCTION

The avifauna of the Kanuku Mountains in southwestern Guyana remains poorly known, despite visits by naturalists to the region since the mid-1800s (Parker et al., 1993). This is particularly true for the Eastern Kanuku Mountains and adjacent lowlands for which no avifaunal assessment has yet been published.

In the Western Kanuku Mountains, a previous expedition by the Rapid Assessment Program in lowland forest, upland forest, and savannahs reported the occurrence of 349 species or about 47% of the birds recorded in the country, highlighting this region's importance with respect to Guyana's bird diversity (Parker et al. 1993). The results of the expeditions to the Eastern Kanuku Mountains presented here document the bird diversity in the Kanuku area, specifically its eastern range along the Lower Kwitaro River.

MATERIALS AND METHODS

Rapid Assessment Program (RAP) Survey

During the expedition, birds were surveyed (primarily by Wiltshire Hinds, with additional observations from Jim Sanderson) at two sites: 1) Pobawau Creek (3°16'3.1"N, 58°46'42.7"W) on 20–24 September 2001 near the confluence of Pobawau Creek and Kwitaro River in occasionally flooded lowland forest, and 2) Cacique Mountain (3°11'29.5"N, 58°48'42.0"W) on 25–29 September 2001, farther up the Kwitaro River about 10 km southwest of Pobawau Creek, in lowland forest (around the camp, 120 m) and hill forest (450 m).

The methods used for compiling the species list included mist-netting, playing back songs recorded on tape, and auditory and visual observations.

Mist-netting was carried out only at Pobawau Creek. On the first day, a total of 12 nets were set along trails in pairs perpendicular to each other. Nets were opened at dawn (530 h), closed at dusk (1730 h) and checked every 90 minutes. The following days the number of nets was reduced to 8 due to longer distances between the nets. The nets used were 6 m in length by 2.6 m in height and 12 m by 2.6 m, and a particular effort was made to mist-net in different habitats, such as in the clear under-story, dense vine tangles, tree fall gaps, and swampy areas.

Song recording was performed using a directional microphone and recorder, primarily at dawn and dusk, but also during the day while walking in the forest. The recorder was equipped with record/playback features, to record unfamiliar bird songs and replay them to attract birds and get a visual identification.

Auditory and visual surveys were emphasized at Cacique Mountain and no mist-netting took place at this site. Observers walked through the forest and floated down the river, observing birds through binoculars and documenting calls. Recording was carried out regularly on these trips.

Davis Finch's Surveys

Davis Finch is a professional birder who has conducted nearly 80 trips to South America, mostly under the auspices of the bird-watching tour company WINGS of which he is a founder and owner. Particularly interested in bird vocalization, Finch has contributed many recordings to the Library of Natural Sounds at Cornell Laboratory of Ornithology, including more than a thousand collected on 18 expeditions to the interior of Guyana. The results of his recent expeditions in and near the RAP study area are included in this report.

Davis Finch's first survey near the RAP study area took place between 10–13 November 1998 near the Kasum base camp (3º11'35"N, 58º48'40"W) on the Kwitaro River. He traveled downriver to the mouth of the Kwitaro River (3º17'25"N, 58º45'01"W), near Pobawau Creek on the last day.

In 2001, Finch surveyed birds between 13–15 November at the same Pobawau Creek camp as the RAP survey and also ascended Cacique Mountain. On both occasions, Finch surveyed birds by auditory and visual identifications. His two surveys produced a list of 210 bird species.

RESULTS

The RAP expedition recorded 176 bird species (Appendix 4). The Pobawau Creek site, with 151 species observed, was slightly more diverse than the Cacique Mountain site (both lowland and hill forests) with 143 species. A total of 35 bird species were observed in the montane forest of Cacique Mountain, but the higher hills and mountains require additional surveys to better document the montane avifauna present in the Eastern Kanukus.

Finch's 1998 survey recorded 178 bird species in the Lower Kwitaro River region near the Kanuku Mountains. His 2001 survey, just two months after the RAP team completed their expedition, recorded 144 species.

Overall, 264 bird species have been recorded for the Eastern Kanuku Mountain region (RAP expedition and Finch's surveys). The most diverse families were tyrant flycatchers (Tyrannidae, 34 species), followed by antbirds (Thamnophilidae, 28 species), tanagers (Thraupidae, 14 species), parrots (Psittacidae, 12 species), woodcreepers (Dendrocolaptidae, 10 species), woodpeckers (Picidae, 10 species), hawks and eagles (Accipitridae, 8 species), and falcons and caracaras (Falconidae, 8 species). Almost 80% of these bird species were associated with lowland forests (210 out of 264 species). The next most favored habitats were rivers and their banks with 62 bird species and scrubs with 57 species (note that a species may be associated with more than one habitat type, see Appendix 4).

Among the birds recorded, 39 bird species are considered uncommon in the Neotropics and two species rare, namely the Harpy Eagle (*Harpia harpyja*) and the Orange-breasted Falcon (*Falco deiroleucus*) (Stotz et al., 1996). Nine species are restricted to the Guianas and adjacent Venezuela and Brazil: Caica Parrot (*Pionopsitta caica*), Green Aracari (*Pteroglossus viridis*), Guianan Toucanet (*Selenidera culik*), Golden-collared Woodpecker (*Veniliornis cassini*), Rufous-bellied Antwren (*Myrmotherula guttata*), Todd's Antwren (*Herpsilochmus stictocephalus*), Painted Tody-Flycatcher (*Todirostrum pictum*), White-throated Manakin (*Corapipo gutturalis*), and White-fronted Manakin (*Lepidothrix serena*). Seven species are further known to be exclusively distributed north of the Amazon: Black Curassow (*Crax alector*), Black-headed Parrot (*Pionites melanocephala*), Black Nunbird (*Monasa atra*), Yellow-billed Jacamar (*Galbula albirostris*), Spot-tailed Antwren (*Herpsilochmus sticturus*), Black-headed Antbird (*Percnostola rufifrons*), and Yellow-throated Flycatcher (*Conopias parva*).

Among the 264 species recorded from the Kwitaro River, 63 had not been observed during the 1993 RAP expedition to the Western Kanukus (Parker et al., 1993). As a result, the avifauna of the Kanukus now numbers 419 species, or 53% of the 786 bird species recorded for Guyana (Braun et al., 2000), indicating that the Kanuku Mountains have a level of avian diversity among the highest in the country. The total number of bird species present in the Kanuku area is likely to be somewhat higher than this figure, given the high proportion (23%) of the species recorded from the Kwitaro lowlands that were new records for the Kanuku area. The Iwokrama reserve is presently considered to have the most diverse avifauna in Guyana, with preliminary results giving numbers comparable to and higher than those we have compiled for the Kanukus (Agro et. al., 1998).

Overall, the lowland forest birds in the Kanuku Mountains are especially well represented with 297 species, or 70% of the 426 birds associated with lowland forest in Guyana. The birds associated with riparian habitats are also well represented with 73 species (70%) out of a total of 104 species occurring in Guyana. The Kanuku Mountains are also rich in birds with small distribution ranges, with 17 out of the 25 species endemic to the Guayana Shield including the nine species mentioned above plus Marail Guan (*Penelope marail*), Rufous-winged Ground-Cuckoo (*Neomorphus rufipennis*), Black-throated Antshrike (*Frederickena viridis*), Brown-bellied Antwren (*Myrmotherula gutturalis*), Rufous-throated Antbird (*Gymnopithys rufigula*), Tiny Tyrant-Manakin (*Tyranneutes virescens*), and Blue-backed Tanager (*Cyanicterus cyanicterus*)—all recorded during the 1993 RAP expedition to the Western Kanuku Mountains. The entire Kanuku Mountains are also home to 10 out of the 15 species distributed north of the Amazon (and present in Guyana): the seven species mentioned above plus Capuchinbird (*Perissocephalus tricolor*), Guianan Cock-of-the-Rock (*Rupicola rupicola*), and Fulvous Shrike-Tanager (*Lanio fulvus*)—also all recorded during the 1993 RAP expedition to the Western Kanukus.

A few remarkable species observed in the Lower Kwitaro River area deserve mention. The Harpy Eagle, *Harpia harpyja*, long known to be reproducing in the Kanuku area (Rettig, 1977, 1978; Parker et al., 1993) was also found present along the Lower Kwitaro River. It would be worthwhile to monitor Harpy Eagle populations to avoid future impacts on this rare species in the Kanuku Mountain region.

The McConnell's Spinetail (*Synallaxis macconnelli*) might be restricted to the Kanuku Mountains area, as the only documented records of this species in Guyana come from the Kwitaro River and Maipaima Creek in the Western Kanukus (where six individuals in the former and one in the latter were taped by Davis Finch).

The Plain-crowned Spinetail (*Synallaxis gujanensis*), although widespread in northern South America and common in many parts of Guyana, is uncommon along the rivers of the interior. In fact, it seems almost restricted to the Kwitaro and Lower Rewa rivers.

The Spot-winged Antshrike (*Pygiptila stellaris*), only recently discovered in Guyana from a few rivers of the interior, has been observed once on the Kwitaro (one individual taped by Davis Finch). The Blackish Antbird (*Cercomacra nigrescens*), also recently discovered in Guyana, is at present only known from the Kwitaro River, appearing to have a very limited range in Guyana.

The Yellow Tyrannulet (*Capsiempis flaveola*), despite its vast range in tropical America is poorly known in the Guayana Shield. Davis Finch recorded this bird along the Takutu River in 2000, but apart from this single occurrence all other observations come from the Kwitaro River (seven individuals altogether) in riverside shrubbery usually including bamboo.

The spectacular Royal Flycatcher (*Onychorhynchus coronatus*) was only mist-netted once during the RAP expedition, at the Pobawau Creek site, and was not observed with any other method, which suggests that the species is uncommon. It also demonstrates that mist-netting, albeit not very productive in terms of number of observations, can be a valuable technique to complement the species list established by usual sound and visual observations.

CONSERVATION AND RESEARCH RECOMMENDATIONS

Documentation of the birds in the Kanuku Mountains and surrounding lowlands is improving, but many species doubtless remain to be recorded. Further efforts must be made to improve our knowledge of this region's avifauna, such as the following:

- Survey the various hill- and mountaintops of the Kanukus to better document birds present in these habitats.

- Conduct a mist-net survey to inventory small cryptic birds in the under-story.

- Study the habitat preferences of uncommon or endemic birds, which potentially require the most urgent protection.

- Monitor the populations of potentially threatened bird species such as the Harpy Eagle (*Harpia harpyja*), which tends to disappear throughout its range; parrots, which may become threatened by the wildlife trade; and game birds.

We also recommend continuing and expanding the existing outreach programs with indigenous communities in the Kanuku Mountains, so as to limit the pressure on birds commonly captured for the wildlife trade (e.g., parrots, birds of prey) by sensitizing the communities to the need to conserve birds in the region and implement hunting practices that are sustainable to prevent local extinction, as 75% of the birds collected for the wildlife trade in Guyana were recorded as captured by the indigenous communities themselves (Edwards, 1992).

CONCLUSION

The Kanuku Mountain region contains one of the highest avifaunal diversities recorded for a single area in Guyana. Several of its birds are rare or uncommon and several others are restricted to the Guianas and adjacent Venezuela and Brazil. The avifauna of this region, in particular in the eastern range, is very close to being in pristine condition. This combination of high diversity, pristine condition, and absence of large human settlements makes the Kanuku Mountains a conservation priority for the country.

LITERATURE CITED

Agro, D., R.S. Ridgely, and L. Joseph. 1998. Preliminary results of surveys of the birds of the Iwokrama Reserve. Abstract for the North American Ornithological Conference, April 6–12. The International Center for Tropical Ecology at the University of Missouri, St. Louis, Missouri.

Braun, M.J., D.W. Finch, M.B. Robbins, and B.K. Schmidt. 2000. A Field Checklist of the Birds of Guyana. Biological Diversity of the Guianas Program Publication 41. Smithsonian Institution, Washington, DC.

Clements, J.F. 2000. Birds of the World: A Checklist. Fifth Edition. Pica Press, Sussex.

Edwards, S.R. 1992. Wild Bird Trade: Perceptions and Management in the Cooperative Republic of Guyana. *In*, Thomsen, J.B., Edwards, S.R., and Mulliken, T.A. Perceptions, Conservation and Management of Wild Birds in Trade. TRAFFIC International, Cambridge, United Kingdom.

Parker, T.A., III, R.B. Foster, L.H. Emmons, P. Freed, A.B. Forsyth, B. Hoffman, and B.D. Gill (eds.). 1993. A biological assessment of the Kanuku Mountain region of southwestern Guyana. RAP Working Papers 5. Conservation International, Washington, DC.

Rettig, N.L. 1977. In quest of the snatcher. Audubon. 79(11): 26–49

Rettig, N.L. 1978. Breeding behavior of the Harpy Eagle (*Harpia harpyja*). Auk. 95(4): 629–643.

Stotz, D.F., J.W. Fitzpatrick, T.A. Parker, III, and D.K. Moskovits. 1996. Neotropical Birds: Ecology and Conservation. University of Chicago Press, Chicago.

Chapter 5

Non-Volant Mammal Survey Results from the Eastern Kanuku Mountains, Lower Kwitaro River, Guyana

Jim Sanderson and Leroy Ignacio

ABSTRACT

This paper presents the results of a survey conducted in the Eastern Kanuku Mountain range along the Lower Kwitaro River from 21–29 September 2001. The purpose of the survey was to assess the biological diversity of non-volant mammals in the region. We used tracks, sound and visual observations, and camera photo-traps to survey for the presence of non-volant mammals. The results of this survey are part of Conservation International's biodiversity corridor project for the Guayana Shield and are intended to identify areas of interest for long-term conservation.

INTRODUCTION

Information on specific local biological diversity is essential to implement effective conservation strategies. Such information, however, is often unknown, incomplete, or unavailable to policymakers. In the case of Guyana, Funk et al. (1999) have already assembled a biodiversity information base that should be used in designing a national system of protected areas. The Kanuku Mountains were mentioned in this publication as one of the five most significant areas for biological conservation, and the Western Kanuku Mountains were the subject of an earlier study by Conservation International's Rapid Assessment Program (Parker et al., 1993). A wide diversity of ecosystems and microclimates have resulted in the Western Kanuku Mountains being identified as containing over 50% of Guyana's avifauna and 70% of its mammalian species. Results from the Rapid Assessment Program were confirmed by a similar survey financed by the European Union and presented to the Guyana government with the recommendation that a national park be established to protect this biodiversity (Agriconsulting, 1993).

As part of its strategy to protect biodiversity, Conservation International has suggested establishing a biological corridor that extends from the Western Kanuku Mountains eastward to Suriname. The Eastern Kanuku Mountains form the larger of the two Kanuku ranges, which together are the central feature of Guyana's southwestern region. Parker et al. (1993) stated, "we think it probable that the Eastern Kanukus (east of the Rupununi River) are somewhat richer in mammal species than the western portion of the range." Since little was known of the biodiversity of the Eastern Kanuku Mountains (Funk et al., 1999), a Rapid Assessment Program survey was undertaken in this area.

The objective of the Rapid Assessment Program in the Eastern Kanuku Mountains was to provide quick, efficient, reliable, and cost effective biodiversity data on this little known region of Guyana to support a regional conservation strategy. We surveyed for non-volant mammals using direct observation, sounds, track identification, and camera photo-traps in the lowland, seasonally inundated, and *terra firma* tropical evergreen forest along the western banks of the Lower Kwitaro River in southwestern Guyana.

MATERIALS AND METHODS

Study Area

We conducted our study from 21–29 September 2001, representing the beginning of the dry season. Our sample sites included Pobawau Creek (3°16'3.1"N, 58°46'42.7"W) and Cacique Mountain (3°11'29.5"N, 58°48'42.0"W) 10 km southwest of Pobawau Creek. Both sites were located on the Kwitaro River in the Rewa River Basin, which is a tributary of the Rupununi River in the larger Essequibo drainage system and within southern Guyana's Region 9. Both sites were approximately 120 m elevation and vegetation was lowland seasonally inundated and *terra firma* evergreen tropical forest. River depth was high for the dry season but dropped rapidly, falling approximately 1.5 m during our brief visit.

Methods

We used direct observation of species, track identification, sound identification, and camera photo-trap records to determine presence of mammalian species in two study areas. Based on camera photo-trap records we computed the Relative Abundance Index (RAI) of each species photographed during each sampling period, defined as 24 hours beginning at 12:00 am, by dividing the number of individuals of a species photographed by the total number of individuals of all species photographed. If multiple photographs of what was thought to be a single individual were recorded during a sampling period (such as sequential photographs of the same individual) then this was counted as a single presence record.

Direct observations and track and sound identification were made during daily excursions from base camp. Surveys were carried out at night using a spotlight. Our colleagues also made opportunistic records and we received additional information through personal communication with our field guides.

Eight CamTrakker photo-traps (CamTrakker Watkinsville, Georgia) were operated simultaneously at each of two study sites. CamTrakker photo-traps are triggered by heat-in-motion. Time between sensor reception and a photograph is 0.6 seconds. Cameras were placed at sites with mammal species evidence; den sites, trails, and feeding stations such as Brazil nut trees were typically chosen for camera placement. Each was located approximately 500 m apart and at least 500 m from base camp.

Camera photo-traps were active continuously. After a photograph was taken a minimum 20 seconds (the minimum setting on a CamTrakker) elapsed before a subsequent photograph could be shot. Ten and nine camera photo-trap periods took place at Pobawau Creek and Cacique Mountain, respectively.

RESULTS

Observations

We observed visually, by sound, or photographed 24 species of non-volant mammals. Because there was no difference in the mammalian fauna between the study sites we have combined the site lists (Appendix 5). The common gray four-eyed opossum (*Philander opossum*) and the common opossum (*Didelphis marsupialis*) were photographed. Other opossums have been recorded but were not found in our survey (Emmons and Feer, 1990). Primates were the most commonly observed and heard mammalian species, and included the common squirrel monkey (*Saimiri sciureus*), brown capuchin monkey (*Cebus apella*), wedge-capped capuchin (*C. olivaceus*), brown bearded saki (*Chiropotes satanus*), red howler monkey (*Alouatta seniculus*), and black spider monkey (*Ateles paniscus*). Golden-handed or Midas tamarin (*Saguinus midas*) and the white-faced saki (*Pithecia pithecia*) were known to occur in the study areas (Duane de Freitas, Ignacio Rufino, Julian James, pers. comm.) but were not observed. The red howler monkeys were heard daily at sunrise.

Kinkajou (*Potos flavus*) were observed and heard at night at Cacique Mountain. The South American coati (*Nasua nasua*) and crab-eating raccoon (*Procyon cancrivorus*) were said to occur in the study areas (Duane de Freitas, Ignacio Rufino, Julian James, pers. comm.). Bush dogs (*Speothos venaticus*) were reported to be present but extremely rare (Jonah Simon, pers. comm.). Three giant otter (*Pteronura brasiliensis*) were observed and filmed on a single occasion at a small inlet on the Kwitaro River downstream from the Pobawau Creek site.

We believe that six felids occur in the study areas; only one, a jaguar (*Panthera onca*) was heard calling. Jaguarundi (*Felis yagourundi*), oncilla (*Leopardus tigrina*), and puma (*F. concolor*) were reported to occur. Margay (*L. wiedii*) and ocelot (*L. pardalis*) were captured on film.

The tracks of gray brocket deer (*Mazama gouazoubira*) and red brocket deer (*M. americana*) were also observed. Capybara (*Hydrochaeris hydrochaeris*) was known to occur in the study area, and was observed on the Rewa River by the RAP team returning to Annai at the end of the expedition. Brazilian tapir (*Tapirus terrestris*) was confirmed at the study sites.

No squirrels, Sciuridae, were observed at either study site. Paca (*Agouti paca*), green acouchi (*Myoprocta exilis*), and red-rumped agouti (*Dasyprocta cristata*) tracks were observed at both study sites; all were phototrapped.

Camera Photo-traps

A total of 51 and 74 photographs were taken at Pobawau Creek and Cacique Mountain, respectively, during 160 camera periods indicating a moderate amount of activity in both areas. Fourteen species of mammals were phototrapped. Additionally, the black curassow (*Crax alector*) and gray-winged trumpeter (*Psophia crepitans*) were also phototrapped. Collared peccary (*Tayassu tajacu*), white-lipped peccary

(*Tayassu pecari*), red-rumped agoutis (*Dasyprocta cristata*), and green acouchi (*Myoprocta exilis*) accounted for 41.4% of all photographs at the two study sites (Table 5.1). The Relative Abundance Index confirmed that prey species were relatively more abundant than predators. It should be noted, however, that we could not estimate the density of any species based on these data.

DISCUSSION

Our results suggest that the full biologically rich assortment of non-volant mammals present in the Western Kanuku Mountains is also found in the Eastern Kanuku Mountains. Moreover, we observed the brown bearded saki (on several occasions) and received confirmation from our guides that the golden-handed or midas tamarin occur in the region, thus confirming Parker et al.'s (1993) suspicions that all eight of Guyana's primates occur in the Eastern Kanuku area.

Our data suggest that agoutis and two species of peccaries were the most abundant herbivores in the lowland, seasonally flooded forests. We also documented the presence of ocelots, margay, and jaguar in the area. According to inhabitants of the local communities, pumas were the most commonly observed carnivores in the region. This perceived abundance may be due in part to their large size, cattle killing habits, or because they encounter people more frequently than jaguars. One local counterpart from Shea Village reported he had observed only three bush dogs in his lifetime, confirming that these carnivores are indeed rare where they occur at all.

Our local counterparts from Shea Village explained that people had previously lived in the higher forested areas close to the Lower Kwitaro River through the 1970s when the balata rubber market collapsed. We were shown balata and mango trees and areas of secondary growth over previously cultivated fields, providing evidence that humans occupied areas along the river that were not seasonally flooded. Shea Village in the Rupununi savannah is the nearest occupied village, located approximately 60 km from our study site.

CONSERVATION RECOMMENDATIONS

Inhabitants of Shea, Maruranau, and Awariwaunau villages in the south and Rewa, Apotari, and Annai villages in the north utilize the Eastern Kanuku Mountains in season for hunting, fishing, and harvesting non-timber forest products. Duane de Freitas of Dadanawa Ranch operates an ecotourism business that organizes an average of two trips per year to the Kwitaro and Rewa Rivers, among others. We believe these activities should continue if protected area status is granted to the Kanuku Mountains and indeed argue that such activities are essential to maintaining this vast biologically rich area.

Conservation International maintains a regional office in Georgetown and a satellite office in Lethem. The Lethem office is building a community outreach program involving 18 local Amerindian villages including Shea, Nappi, and Shulinab located near the Kanuku Mountains. We plan to visit four or five of these villages which have expressed high interest in working cooperatively in the camera-trapping program. We will provide camera photo-trapping instruction and assistance to deploy cameras photo-traps in each village. Every ten weeks film and batteries in the photo-traps

Table 5.1. Relative Abundance Index (in %) for Pobawau Creek and Cacique Mountain sites.

Scientific Name	Common Name	Relative Abundance (in %)
Leopardus wiedi	Margay	1.5
Mazama gouazoubira	Gray brocket deer	1.5
Priodontes maximus	Giant armadillo	1.5
Didelphis marsupialis	Common opossum	3.1
Dasypus kappleri	Great long-nosed armadillo	3.1
Leopardus pardalis	Ocelot	4.6
Mazama americana	Red brocket deer	4.6
Philander opossum	Common gray four-eyed opossum	6.2
Procechimys guyanensis	Spiny rat	9.2
Agouti paca	Paca	9.2
Dasyprocta cristata	Red-rumped agouti	13.8
Myoprocta exilis	Green acouchi	13.8
Tayassu tajacu	Collared peccary	13.8
Tayassu pecari	White-lipped peccary	13.8

will be replaced, and camera photo-traps will be relocated once every year. All photographs will be made available to each village and once a year all participants will meet to compare and discuss results. In this way a sustained long-term camera photo-trapping research and outreach program can be undertaken that provides villagers and scientists with continuous information on local wildlife resources enabling more informed conservation actions.

ACKNOWLEDGMENTS

The authors wish to thank Duane de Freitas for additional information on local wildlife. We also thank Jonah Simon, Julian James, and Ignacio Rufino of Shea Village for their assistance.

LITERATURE CITED

Agriconsulting. 1993. Preparatory study for the creation of a protected area in the Kanuku Mountains region of Guyana. Unpublished Report. European Development Fund, Rome, Italy.

Emmons, L.H. and F. Feer. 1990. Neotropical Rainforest Mammals. University of Chicago Press, Chicago.

Funk, V., M.F. Zermoglio, and N. Nasir. 1999. Testing the use of specimen collection data and GIS in biodiversity exploration and conservation decision-making in Guyana. Biodiversity and Conservation. 8: 727–751.

Lim, B.K and Z. Norman. 2002. Rapid Assessment of Small Mammals in the Eastern Kanuku Mountains, Lower Kwitaro River Area, Guyana. *In*, Montambault, J.R. and O. Missa (eds.). A Biodiversity Assessment of the Eastern Kanuku Mountains, Lower Kwitaro River Area, Guyana. RAP Bulletin of Biological Assessment 26. Conservation International, Washington, DC.

Nowak, R.M. 1991. Walker's Mammals of the World. Fifth edition. Johns Hopkins University Press.

Parker, T.A., III, R.B. Foster, L.H. Emmons, P. Freed, A.B. Forsyth, B. Hoffman, and B.D. Gill (eds.). 1993. A biological assessment of the Kanuku Mountain region of southwestern Guyana. RAP Working Papers 5. Conservation International, Washington, DC.

Chapter 6

Rapid Assessment of Small Mammals in the Eastern Kanuku Mountains, Lower Kwitaro River Area, Guyana

Burton K. Lim and Zacharias Norman

ABSTRACT

Two sites in the Eastern Kanuku Mountains along the Kwitaro River were surveyed for small mammals using mesh mist nets for bats and folding box-style traps for rodents and marsupials between 20–28 September 2001. Thirty-two species of small mammals were documented including 27 species of bats, 3 species of rats, and 2 species of opossums. For bats, two species of fruit-eaters (*Artibeus planirostris* and *A. obscurus*) were the most abundant and represented almost half (114) of the 234 bat captures. We added three species that are new to the fauna of the Kanuku Mountains, and five species documented for the first time from the eastern region. Incorporating previous studies, there are 155 species of mammals presently known from the Kanuku Mountains region, including the forested Kanuku Mountains south to the savannahs of Aishalton, which represents almost 70% of the known species diversity for Guyana.

INTRODUCTION

Until recently, the small mammal biodiversity in Guyana had been largely unknown and was inferred primarily from earlier studies conducted in Venezuela (e.g., Handley, 1976; Linares, 1998) and Suriname (e.g., Husson, 1978). Small mammals (bats, rats, and opossums), however, represent approximately 75% of the mammal species diversity in the country (Engstrom and Lim, 2002). Furthermore, this faunal assemblage fills an important ecological role in the environment by acting as seed dispersers, flower pollinators, insect controllers, and prey species.

Guyana is in the process of establishing a National Protected Areas System as outlined in the Convention on Biological Diversity that resulted from the 1992 United Nations Conference on Environment and Development. The Kanuku Mountains in southwestern Guyana have been identified as an area worthy of conservation by two independent studies (Agriconsulting, 1993; Parker et al., 1993). The state of knowledge of mammal diversity in this region, however, is erratic, ranging from a reported 60 species (Agriconsulting, 1993) to 150, based predominately on unverified specimens at the Royal Ontario Museum (ROM) supplemented with data from a previous RAP expedition to the Western Kanuku Mountains (Parker et al., 1993). The objectives of this small mammal project were to survey the eastern region, verify the species documented from the area, and update the conservation status of the Kanuku Mountains.

MATERIALS AND METHODS

Small mammals were surveyed primarily by using mesh mist nets and live box traps. Bats were captured with three sizes of nets (6 m in length by 2.6 m in height, 12 m by 2.6 m, and 30.5 m by 9.1 m) set in a variety of forest microhabitats. On any given night, up to 24 smaller nets were set in the forest under-story to a height of 3 m above the ground. They were placed where bats might be expected in high concentrations because of roost sites or food resources such as swamps, rocky areas, dry creek beds, rivers, vine tangles, tree hollows, fruiting or flowering trees, and upland forest. One large net was set along a dry intermittent stream at the base of a hill and raised 20 m into the canopy. Nets were typically opened from dusk to dawn unless there was rain or if some nets set near roost sites or fruiting trees captured high numbers of one or a few species. In the latter case, the nets were closed at 2200 h. Netting effort was calculated as the number of hours of net surface-area deployed. For example, a small net opened all night yielded 187.2 m²h of effort (6 m by 2.6 m by 12 h).

Non-flying small mammals, such as rats and opossums, were captured using two types of live, folding box-style traps. We used two sizes of aluminum Sherman traps (used to target rats and mouse opossums) with a length by width by height of 23 cm by 8 cm by 9 cm, and 35 cm by 12 cm by 14 cm. They were set on the ground near burrows or along foraging runways near the bases of large trees, along tree falls, and in rocky areas in upland forest and swamps for terrestrial animals. Some traps were also set on vines and low branches to sample the arboreal population. Up to 80 small and 40 large Sherman traps were set on any given day in line transects with traps spaced approximately 5 m apart. Two sizes of wire Tomahawk traps (used to target larger opossums) were used, measuring 60 cm by 24 cm by 24 cm, and 100 cm by 50 cm by 50 cm. A maximum of seven Tomahawk traps were strategically set on the ground near large trees with vines leading up to the crowns to increase the chances of encountering animals. Trapping effort was calculated as the number of traps set each night.

Voucher specimens were prepared as dried skins with skulls and skeletons, or as whole animals fixed in 10% formalin with long-term storage in 70% ethanol. Tissue samples of liver, heart, kidney, and spleen were frozen in the field with liquid nitrogen and later stored in a -80° C ultra-cold freezer. Specimens will be deposited at the Centre for the Study of Biological Diversity at the University of Guyana and the Royal Ontario Museum, Canada.

The small mammal survey was conducted at two main localities situated on the western bank of the Kwitaro River within the area designated as the Kanuku Mountains region described in Parker et al. (1993). This region encompassed approximately 15,000 square kilometers of forest and savannah with elevations to almost 1000 m. The first collecting site on this RAP expedition was at the mouth of Pobawau Creek (3°16'3.1"N, 58°46'42.7"W, 120 m elevation). Camp was situated at the apex of a gradually broadening portion of forest between the two bodies of water. Nets and traps were set within 500 m of camp from 20–22 September 2001 (traps were kept out for an extra day on 23 September). The habitat in this area included dry seasonally flooded forest, vine tangles, river edges, and upland forest. The second site was located at a landing on the Kwitaro River near to Cacique Mountain (3°11'29.5"N, 58°48'42.0"W, 120 m), an outlying peak of the Kanuku Mountain range. Nets and traps were set within 1 km of

camp, and reached the lower slopes of the mountain. Microhabitats, including dry creek beds, swamp, rocky areas, and upland forest, were sampled for 5 nights from 24–28 September 2001.

Simpson's index $[-\ln \sum ((n^2 - n) / (N^2 - N))$, where n is the number of individuals for a species and N is the total number of individuals for all species] was calculated to assess the species diversity between the two sites and for comparison with Iwokrama Forest, which is the only other locality in Guyana with comparable data (Engstrom and Lim, submitted). Chao's estimator $[S^* = S_{obs} + (a^2 / 2b)$, where S^* is the expected number of species, S_{obs} is actual number of species observed, a is the number of species collected once, and b is the number of species collected twice] was used to predict the total number of species during the project.

RESULTS AND DISCUSSION

Species Diversity

A total of 32 species of small mammals were documented from two study sites in the Eastern Kanuku Mountains (Tables 6.1–6.3). Colonies of proboscis-nosed river bats (*Rhynchonycteris naso*) were commonly seen roosting on tree trunks overhanging the Kwitaro River but none were captured in nets. The other 31 species were documented by 245 individuals, of which 157 were kept as voucher specimens. The taxonomic composition included two species of marsupials represented by two individuals, three species of rats (9 individuals), and 26 species of bats represented by 234 individuals, 88 of which were released at the point of capture. Three other species of small mammals were observed but positive identifications were not possible. At least two medium-sized dark-colored sheath-tailed bats (Family Emballonuridae) were disturbed in the daytime from their roost in the hollow of a rotted tree-fall across a dry intermittent creek bed near the camp at Cacique Mountain. An arboreal rat and mouse opossum were also seen at night in trees by the boat landing at Cacique Mountain.

For bats, which have the most complete data, Simpson's index of diversity was 1.93. This diversity index is relatively low, only one of the nine collecting sites at Iwokrama Forest in central Guyana had a lower value, but sampling at these localities ranged from 6 to 15 days (Engstrom and Lim, submitted) compared with 3 and 5 days at the two sites in this study. Species were still being discovered on the last day of collecting at both sites, thus final comparisons of diversity indices at this time are premature but give a rough gauge of sampling progress. The last collecting night was the most productive of the trip with 14 species of bats and 39 individuals captured.

Collecting Effort

Trapping effort included 43 Tomahawk trap nights (Table 6.1), which yielded two species of marsupials, the gray four-eyed opossum (*Philander opossum*) and the commonstink

Table 6.1. Non-flying small mammal species captured by wire Tomahawk traps in the Eastern Kanuku Mountains along the Lower Kwitaro River with associated effort and catch per unit effort.

Species	Common Name	Total	Pobawau Creek	Cacique Mountain
Didelphis marsupialis	Common opossum	1	1	0
Philander opossum	Gray four-eyed opossum	1	1	0
Proechimys cuvieri	Cuvier's terrestrial spiny rat	1	1	0
Total		**3**	**3**	**0**
Unit Effort (Trap Nights)		**43**	**17**	**26**
Catch Per Unit Effort		**0.0698**	**0.1765**	**0**

Table 6.2. Non-flying small mammal species captured by aluminum Sherman traps in the Eastern Kanuku Mountains along the Lower Kwitaro River with associated effort and catch per unit effort.

Species	Common Name	Total	Pobawau Creek	Cacique Mountain
Oryzomys megacephalus	Large-headed terrestrial rice rat	4	0	4
Proechimys cuvieri	Cuvier's terrestrial spiny rat	1	1	0
Proechimys guyannensis	Guiana terrestrial spiny rat	3	0	3
Total		**8**	**1**	**7**
Unit Effort (Trap Nights)		**750**	**157**	**593**
Catch Per Unit Effort		**0.0107**	**0.0064**	**0.0118**

opossum (*Didelphis marsupialis*), and one terrestrial spiny rat (*Proechimys cuvieri*). Catch per unit effort from the Tomahawk traps was therefore 7% successful. There were 750 Sherman trap nights, which resulted in the capture of eight rats including two species of terrestrial spiny rats (*P. cuvieri* and *P. guyannensis*) and one species of terrestrial rice rat (*Oryzomys megacephalus*), with a success rate of 1% (Table 6.2). The overall success rate for non-flying small mammals was 1.4%, which was almost triple the rate (0.5%) reported from Iwokrama Forest (Engstrom and Lim, submitted).

Trapping effort for bats totaled 57,601.8 m^2h and yielded 234 individuals for a success rate of 0.4% per m^2h of net (Table 6.3). In more practical terms, on average, 1.3 bats were caught per night per long net (12 m by 2.6 m) set. This capture rate was almost identical to that reported from Iwokrama Forest (Engstrom and Lim, submitted). Between sites, the success rate at Cacique Mountain (0.44%) was one-third better than that at Pobawau Creek (0.33%).

Species Abundance

The two most abundant bats encountered during this survey were fruit-eating bats (*Artibeus planirostris* and *A. obscurus*) that feed primarily on figs. They accounted for almost half (49%) of the total captures (Table 6.3). A similar ecological dominance of the fruit-eating trophic guild was also observed

at Iwokrama Forest; however, another fig-eating specialist (*A. lituratus*) was the most dominant at that site (Lim and Engstrom, 2001b). This species was only moderately abundant during our study. The overall rank order of species abundance, however, was heavily influenced by the higher number of captures from Cacique Mountain as indicated by the identical top four species for both these totals. This trend was not apparent in the bat community structure at Pobawau Creek where the usually common insect-feeding mustached bat (*Pteronotus parnellii*) was caught only once. At this site, 7 of the 17 species documented were captured once, whereas at Cacique Mountain 10 of 22 documented species were caught one time. For the combined sites, 10 of the 26 species were represented by one individual captured during the study period. Data on non-flying small mammals were too sparse to make any comments on relative abundance for this group.

Trophic Structure

For bats, the fruit-eating trophic guild was the most dominant group in terms of relative abundance representing over two-thirds (69%) of the total capture, and also for species diversity with just over half (54%) of the number of species collected. None of the other trophic guilds were represented by more than three species: aerial insectivores (*Pteronotus parnellii, Myotis nigricans,* and *M. riparius*), gleaning insectivores (*Tonatia silvicola* and *Mimon crenulatum*), carnivores

(*Trachops cirrhosus* and *Chrotopterus auritus*), omnivores (*Phyllostomus elongatus* and *P. hastatus*), nectar-feeders (*Lonchophylla thomasi* and *Lionycteris spurrelli*), and blood-feeders (*Desmodus rotundus*). In terms of relative abundance, the aerial insectivores accounted for 14% of the captures, and carnivorous species were represented by 7% of the captures. The remaining trophic guilds accounted for less than 5% of the relative abundance.

Although the survey period was of short duration, the trophic structure of the bat community suggests that there were many trees fruiting, especially large fig trees, the fruits of which are heavily eaten by *Artibeus* spp. Of note also was

the relatively significant number of carnivorous bats caught indicating that small vertebrate prey such as frogs and lizards were probably relatively abundant. Two vampire bats were caught suggesting that there were reasonable populations of large mammals in the area such as tapir and deer. Although they can be quite abundant in inhabited areas where domesticated livestock provide a readily available food source, vampire bats are typically uncommon to rare in deep forest. Several nectar-feeding bats were also caught indicating that there were flowering trees in the area. Usually, nectar-feeding bats are relatively difficult to catch because they are small, fluttering flyers that are not easily captured in nets. This is

Table 6.3. Rank order of bat species captured by netting in the Eastern Kanuku Mountains at two sites on the Lower Kwitaro River with associated effort and catch per unit effort.

Rank	Species	Common Name	Total Captures	Pobawau Creek	Cacique Mountain
1	*Artibeus planirostris*	Large fruit-eating bat	61	6	55
2	*Artibeus obscurus*	Sooty fruit-eating bat	53	16	37
3	*Pteronotus parnellii*	Greater mustached bat	29	1	28
4	*Carollia perspicillata*	Common short-tailed fruit bat	19	9	10
5	*Trachops cirrhosus*	Frog-eating bat	10	7	3
6-7	*Artibeus lituratus*	Greater fruit-eating bat	9	3	6
6-7	*Tonatia silvicola*	White-throated round-eared bat	9	2	7
8-9	*Chrotopterus auritus*	Woolly bat	6	1	5
8-9	*Lonchophylla thomasi*	Spear-nosed long-tongued bat	6	1	5
10	*Phyllostomus elongatus*	Dark spear-nosed bat	5	2	3
11-12	*Platyrrhinus helleri*	Heller's white-lined fruit bat	4	3	1
11-12	*Rhinophylla pumilio*	Little fruit bat	4	1	3
13	*Carollia brevicauda*	Silky short-tailed fruit bat	3	0	3
14-16	*Myotis riparius*	Riparian myotis	2	2	0
14-16	*Artibeus glaucus*	Little fruit-eating bat	2	1	1
14-16	*Desmodus rotundus*	Common vampire bat	2	2	0
17-26	*Artibeus gnomus*	Dwarf fruit-eating bat	1	0	1
17-26	*Mesophylla macconnelli*	Macconnell's bat	1	0	1
17-26	*Mimon crenulatum*	Striped hairy-nosed bat	1	0	1
17-26	*Myotis nigricans*	Black myotis	1	1	0
17-26	*Lionycteris spurrelli*	Chestnut long-tongued bat	1	0	1
17-26	*Phyllostomus hastatus*	Greater spear-nosed bat	1	0	1
17-26	*Platyrrhinus brachycephalus*	Short-headed white-lined bat	1	1	0
17-26	*Sturnira tildae*	Greater yellow-shouldered bat	1	0	1
17-26	*Uroderma bilobatum*	Common tent-making bat	1	0	1
17-26	*Vampyressa bidens*	Common yellow-eared bat	1	0	1
Total			234	59	175
Netting Effort (M²h)			57,601.8	17,971.2	39630.6
Catch Per Unit Effort			0.0041	0.0033	0.0044
Simpson's Index			1.93	2.14	1.75

anecdotal information gleaned from our survey data warranting further investigation.

Spatial Variation

At Pobawau Creek, there were 21 species of small mammals documented composed of two species of opossums, one species of rat, and 18 species of bats (the proboscis-nosed river bat was only observed). There were 25 species recorded from Cacique Mountain including two species of rats and 23 species of bats (the proboscis-nosed river bat was also only observed here).

It was interesting to note that both species of marsupials and Cuvier's terrestrial spiny rat were caught only at Pobawau Creek, whereas the Guianan terrestrial spiny rat and the terrestrial rice rat were caught only at Cacique Mountain (Tables 6.1 and 6.2). We caution against any further interpretation because of the small sample size and relatively short sampling period. More surveying would be needed to test whether this was an artifact of sampling or a true distinction in species distribution associated with habitat differences. Of additional note was that all four individuals of rice rats were caught in swampy habitat or surrounding areas at Cacique Mountain, whereas the three individuals of the Guianan terrestrial spiny rat were all caught on higher ground presumably less prone to flooding. Again, more study is required to verify the existence of this habitat preference.

Seventeen species of bats were captured at Pobawau Creek and 22 species at Cacique Mountain (Table 6.3). Half (13) of the 26 bat species captured were found at both localities. Four species were netted at only Pobawau Creek, and nine species at only Cacique Mountain. Simpson's diversity index, however, was higher (2.14) for Pobawau Creek than that for Cacique Mountain (1.75). The species evenness at Cacique Mountain was skewed by high captures of two species of fruit-eating bats in a few nets that were set near fruiting trees as indicated by partially eaten figs (*Ficus* spp.) that were found below the nets. This highlights one of the drawbacks of diversity indices because a disproportional probability of capture will lower the value even though the absolute diversity may actually be high. Ideally, one to two weeks of collecting at each site would result in data more conducive for comparative purposes and robustness.

Completeness of Survey

As expected from only an eight-day study, the survey of small mammals is far from complete. This is evident from the species accumulation curve, which is still rising sharply with the discovery of five additional species on the last day of collecting (Figure 6.1). Other methods of estimating total number of species also reflect the short survey period and fall short of predictive power. For example, Chao's estimator predicts only 18 more species for a total small mammal diversity of 50 species. From the first RAP expedition, a list of 127 bat species was compiled for the region from the Eastern Kanuku Mountains south to Aishalton (Parker et al., 1993). Although this area includes the southern Rupununi savan-

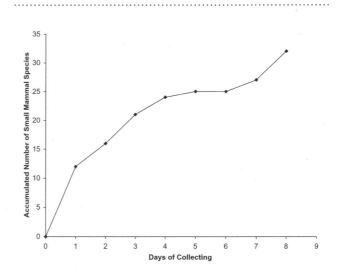

Figure 6.1. Species accumulation curve for small mammals based on the number of days of collecting along the Lower Kwitaro River in the Eastern Kanuku Mountains of Guyana from 20–28 September 2001.

nahs, there are only two species of bats (*Glossophaga longirostris* and *Molossus pretiosus*) that are considered primarily non-forest species (Lim and Engstrom, 2001a). There are, however, several more mammal species that are commonly associated with savannah but are also found to lesser degrees in forested areas, such as the white-tailed deer.

History of Mammal Collecting in the Kanuku Mountains

John J. Quelch carried out the first scientific mammal collection in the Kanuku Mountains in 1900, depositing 28 species in the British Museum of Natural History (Thomas, 1901). Not until 1961 was a second mammal expedition conducted at Nappi Creek in the Western Kanuku Mountains by R. L. Peterson of the Royal Ontario Museum (ROM), Canada. From this initial exploratory trip, he established a field collecting collaboration from 1962 to 1972 with S. E. Brock and J. Marques at Dadanawa Ranch, resulting in a comprehensive collection of mammals from the surrounding savannahs and forest being deposited at the ROM. Stan Brock also collected some specimens for the Smithsonian Institution in Washington, DC. In 1990, the ROM renewed its fieldwork in Guyana with a short four-day trip to Kuma River in the Western Kanuku Mountains by M. D. Engstrom, F. A. Reid, and B. K. Lim. Most recently, L. H. Emmons revisited the Western Kanuku Mountains in 1993 at Mapaima Creek on a RAP expedition (Parker et al., 1993).

Bat Diversity

As presently understood, bats comprise over half (54%) of the mammal species diversity in Guyana (Engstrom and Lim, 2002). This proportion is similar to that currently documented from the Kanuku Mountain region (89 of 154), as amended below. One species of bat (*Chiroderma salvini*) listed from the Eastern Kanuku Mountains in Parker et al. (1993) was based on a misidentification and was subsequently proven not to exist in this region or Guyana. From the current RAP expedition to the Eastern Kanuku Mountains, we documented four additional bat records to this region (*Artibeus glaucus, Mesophylla macconnelli, Platyrrhinus brachycephalus,* and *Myotis riparius*). In addition, three more bat species have been newly documented from this region (*Saccopteryx gymnura* in the east, *Micronycteris homezi* in the east, and *Promops centralis* in the west) (Lim and Engstrom, 2001a). This brings the bat fauna for the Kanuku Mountains region to 89 species, which surpasses the 86 species reported from Iwokrama Forest. This is not to diminish the biodiversity of Iwokrama Forest because it is one-fourth the area of the Kanuku Mountains region and does not include savannah habitats. Pending designation of this region as a national park or biosphere reserve and determination of the exact boundaries and habitats including forest and savannah, the Kanuku Mountains region could harbor the highest known bat species diversity within any protected area in the world. The only addendums to this are that the results of the revisions from the current taxonomic review of the earlier collections may alter the species number; and in actuality the western Amazon probably has the highest bat and mammal diversity in the world (see Reid et al., 2000) but areas such as Manu Biosphere Reserve in Peru have not been thoroughly sampled for some groups including the high flying aerial insectivorous bats (Voss and Emmons, 1996; Lim and Engstrom, 2001a). The unpublished bat species list for Manu is currently at 94 (B. D. Patterson, pers. comm.).

Mammal Diversity in the Kanuku Mountains

A species list for this area was compiled from the first RAP expedition to the Kanuku Mountains (Parker et al., 1993), which was heavily based on the Brock and Marques collections, but these have not been taxonomically reviewed since their deposition at the ROM in the 1960's. A total of 150 species of mammals were listed for the forested Kanuku Mountains south to the savannahs of Aishalton. The Eastern Kanuku Mountains had 86 species, the Western Kanuku Mountains had 127 species, and the savannahs had 22 species. In addition to the new bat records reported above, we document one new species of rat to the Kanuku Mountains region (*Proechimys guyannensis*). This brings the total mammalian fauna known from the Western Kanuku Mountains to 132, Eastern Kanuku Mountains to 89, and the whole region to 155 species.

Since the prior estimate of 175–190 species (Parker et al., 1993), the total number of mammal species from Guyana has been revised to at least 225, most of which are represented primarily by museum voucher specimens (Engstrom and Lim, 2002). A large portion of this increase in species diversity has been within bats, where 24 new country records were reported recently, based largely on intensive faunal surveys in Iwokrama Forest (Lim et al., 1999; Lim and Engstrom, 2001a). Similar field methodology, including more systematic deployment of canopy nets for bats (Engstrom and Lim, submitted), will undoubtedly result in more species being documented from the Kanuku Mountains. Based on our current knowledge, the Kanuku Mountains region harbor 70% of the mammal fauna of Guyana, however, the actual value is probably around 80% because we anticipate that species discovery will be higher for this specific area as a subset than for the country as a whole.

CONCLUSIONS

* The Kanuku Mountains region harbors a high proportion (70% documented with an actual value probably closer to 80%) of the mammalian fauna of Guyana. By conserving this region, a large segment of the mammal biodiversity will be protected.

* Bats comprise over half (57%) of the mammal species diversity in the Kanuku Mountains region and fill an important role in the ecosystem as seed dispersers, flower pollinators, and insect-population controllers. At 89 bat species, this will be the highest reported diversity of bats in any protected area in the world. This distinction is achieved only with the inclusion of a wide range of habitat types found within the region including savannah, riverine forest, and montane areas.

* The bat community structure recorded suggests that the ecosystem of the Kanuku Mountains region is quite variable and healthy. Indicators included a high relative abundance of fruit-eating bats, especially fig specialists (*Artibeus* spp.); the presence of nectar-feeders indicating that trees were flowering; vampire bats suggesting that large mammals were in the area, and the presence of carnivorous species, indicating that good populations of small vertebrates must exist.

* Three species of small mammals were recorded for the first time from the Kanuku Mountains and five species from the eastern region as a result of this eight-day RAP survey. This indicates that there is still more species diversity to be discovered and documented. More detailed and longer term studies beyond this RAP are needed to obtain a full or nearly complete inventory of mammals, and to investigate several sampling patterns that arose from this survey such as the association of species distributions with habitat differences.

CONSERVATION RECOMMENDATIONS

- Designating the Kanuku Mountains region as some type of protected area (e.g., National Park or Nature Reserve) has the potential for conserving a large proportion of the mammal biodiversity found in Guyana depending on the exact delineation for the boundaries and establishment of a sound management program.

- The protected area must incorporate both of the major habitat types in the region including forest and savannah. In addition, a variety of microhabitats within these biomes should be present such as gallery forest, swamps, and montane areas. This mix will ensure the highest and healthiest mammal species diversity but it is also dependent on the eventual designation of protected area category and its management and use.

- The current state of biological knowledge for mammal diversity in the Kanuku Mountain region is relatively good compared to other faunal groups, primarily because of the earlier long-term involvement of Dadanawa Ranch collecting in the savannahs and adjacent forest in the south Rupununi. Despite this sound taxonomic base, in only eight days we collected five species of small mammals previously unknown from the eastern region. This indicates that more biodiversity survey work is needed to document the actual diversity, abundance, and local distribution of mammals that live in this area.

- Funds should be sought to establish a ranger-training program composed of local inhabitants and a field station similar to what has been implemented in Iwokrama Forest in central Guyana. After acquiring the proper skill sets, they would be responsible for a biological inventory of the protected area and monitoring programs to track the status of biodiversity and the environment. In particular, attention should be focused on the management of species susceptible to over-hunting. This should be done with long-term partnerships between international institutions in collaboration with the University of Guyana and local communities.

- Combining research-based opportunities with education-based ecotourism can generate revenue. Although bird watching and primates will still probably be the main attractions, other alternative nature activities can be easily started with minimal training such as nocturnal bat netting and the explanation of this group's diversity and importance to natural rainforest succession. Environmental conservation may not be self sufficient, but programs that assist the local economy will undoubtedly ease pressure and increase additional private and public contributions and aid.

ACKNOWLEDGMENTS

We would like to thank Mark Engstrom and Graham Watkins for commenting on an earlier draft of the manuscript. Thanks also to the personnel from Dadanawa and Karanambo in the Rupununi Savannahs for field assistance and transportation. The logistical arrangements were greatly appreciated from the staff at the CI Guyana and Washington offices. This is contribution number 250 from the Centre for Biodiversity and Conservation Biology at the Royal Ontario Museum.

LITERATURE CITED

Agriconsulting. 1993. Preparatory study for the creation of a protected area in the Kanuku Mountains region of Guyana. Unpublished Report. European Development Fund, Rome, Italy.

Engstrom, M.D. and B.K. Lim. 2002. Mamíferos de Guyana. *In*, Ceballos, G. and J. Simonetti (eds.). Diversidad y conservación de los mamíferos de Latinoamérica. Fondo de Cultura Económica-UNAM, México, DF, *in press*.

Engstrom, M.D. and B.K. Lim. The mammals of the Iwokrama Forest. *In*, Joseph, L., G.G. Watkins, and M.D. Engstrom (eds.). The fauna and flora of the Iwokrama Forest. Proceedings of the Academy of Natural Sciences of Philadelphia, *submitted*.

Handley, C.O., Jr. 1976. Mammals of the Smithsonian Venezuelan project. Brigham Young University Science Bulletin, Biological Series. 20(5): 1–91.

Husson, A. M. 1978. The mammals of Suriname. Zoölogische Monographieën. Rijksmuseum van Natuurlijke, Historie, 2: 1–569.

Lim, B.K., M.D. Engstrom, R.M. Timm, R.P. Anderson, and L.C. Watson. 1999. First records of 10 bat species in Guyana and comments on the diversity of bats in Iwokrama Forest. Acta Chiropterologica. 1: 179–190.

Lim, B. and M. Engstrom. 2001a. Species diversity of bats (Mammalia: Chiroptera) in Iwokrama Forest, Guyana, and the Guianan subregion: implications for conservation. Biodiversity and Conservation. 10: 613–657.

Lim, B.K. and M.D. Engstrom. 2001b. Bat community structure at Iwokrama Forest, Guyana. Journal of Tropical Ecology. 17: 647–665.

Linares, O.J. 1998. Mamíferos de Venezuela. Sociedad Conservacionista Audubon de Venezuela, Caracas, 691 pp.

Parker, T.A., III, R.B. Foster, L.H. Emmons, P. Freed, A.B. Forsyth, B. Hoffman, and B.D. Gill (eds.). 1993. A biological assessment of the Kanuku Mountain region of southwestern Guyana. RAP Working Papers 5. Conservation International, Washington, DC.

Reid, F.A., M.D. Engstrom, and B.K. Lim. 2000. Noteworthy records of bats from Ecuador. Acta Chiropterologica. 2:37–52.

Thomas, O. 1901. On a collection of mammals from the Kanuku Mountains, British Guiana. Annals and Magazine of Natural History 7(7): 139–154.

Voss, R.S. and L.H. Emmons. 1996. Mammalian diversity in Neotropical lowland rainforests: a preliminary assessment. Bulletin of the American Museum of Natural History. 230: 1–115.

Gazetteer

The following is a description of collection sites referenced in this publication. All sites are located in the Rewa River basin, which is a tributary of the Rupununi River in the larger Essequibo drainage system. The team traveled by truck to the village Annai, and then continued up the Rupununi, Rewa, and Kwitaro Rivers progressively over a four-day boat journey. The camps were made with the assistance of four indigenous community members from Shea village, located another day's boat trip up the Kwitaro River and approximately 42 km over land. The area was very remote; no people unconnected to the RAP expedition were encountered after the first day of travel.

Pobawau Creek
20–24 September 2001
3°16'3.1"N, 58°46'42.7"W
120 m elevation
Located on the Lower Kwitaro River at the mouth of Pobawau Creek. Camp was situated at the apex of a gradually broadening portion of forest between the two bodies of water. The habitat in this area included dry seasonally flooded forest, vine tangles, river edges, and upland forest.

Cacique Mountain
25–29 September 2001
3°11'29.5"N, 58°48'42.0"W
120 m elevation (at river); 450 m elevation (at highest peak surveyed)
Located approximately 10 km up the Kwitaro River from the Pobawau Creek site. Cacique Mountain is an outlying peak of the Kanuku Mountain range. Microhabitats included dry creek beds, swamp, rocky areas, and upland forest.

Corona Falls (Fishes and Water Quality only)
3°11'35.0"N, 58°48'39.6"W
Located approximately two hours up the Rewa River from the Kwitaro River mouth at Corona Falls rapids. Rapids contained abundant Podostemaceae plants, and an unnamed forest creek was sampled.

Appendices

Appendix 1

Students and instructors participating in the 2001 Guyana RAP Training at Tropenbos West Pibiri Creek Ecological Station, Mabura Hill Township

Eustace Alexander

Students

Aiesha Williams
Environmental Protection Agency (EPA)

Christopher Chin
Iwokrama International Centre

Damian Fernandes
Iwokrama International Centre

Deirdre Jafferally
Iwokrama International Centre

Hemchandranauth Sambhu
Environmental Protection Agency (EPA)

Iwan Derveld
Conservation International–Suriname

Jacky Sutrisno Mitro
STINASU–Suriname

Jan Hendrik Meijer
Conservation International–Suriname

Julian Pillay
Tropenbos–Guyana Programme

Kathleen Fredericks
Iwokrama International Centre

Lindsford La Goudou
Iwokrama International Centre

Marijem Djosetro
STINASU–Suriname

Romeo de Freitas
Guyana Marine Turtle Conservation Society

Students and Field Assistants

Corletta Toney

Justin de Freitas
Dadanawa Ranch

Leroy Ignacio
Rupununi Weavers Society

Wiltshire Hinds
Centre for Study of Biodiversity/
Environmental Protection Agency (EPA)

Zacharias Norman
Iwokrama International Centre

Scientists and Instructors

Burton K. Lim
Royal Ontario Museum–Canada

Jim Sanderson
Conservation International–Washington

Olivier Missa
Conservation International–Washington

Peter Hoke
Conservation International–Washington

Wilmer Díaz
Jardín Botánico del Orinoco–Venezuela

Coordinators

Eustace Alexander
Conservation International–Guyana

Jensen R. Montambault
Conservation International–Washington

Appendix 2

Preliminary list of plants observed or collected during the RAP expedition to the Eastern Kanuku Mountains, Guyana with site and habitat information

Wilmer Díaz

Note: This list reflects preliminary results made in the field, pending final identification, and should be used with caution. Preliminary determinations by the author, unless otherwise noted. Species known to exist only in regions other than the Guayana Shield are indicated with a "?" behind the species name and require further determination. New records are compared to Hoke (2002). See Chapter 2 for reference and discussion. * indicates specimen collected in the rapids of the Corona Falls, Rewa River site.

Collection Locality Habitats
MF – Mixed forest
MFCM – Mixed forest in Cacique Mountain valley
MFS – Mixed forest on Cacique Mountain slope
PC – Along Pobowau Creek
R – Along Kwitaro River
RM – River meander successional vegetation
SFM – Seasonally flooded mixed forest
SV – Shrubby vegetation on top of Cacique Mountain
C – Collected
O – Observed

Determiner
WD – Wilmer Díaz
ES – Elio Sanoja, Universidad Nacional Experimental de Guayana (Undergraduate, Upata; Graduate, Puerto Ordaz, Edo. Bolívar) and Herabario Regional de Guayana (Research Associate)
AL – Angelina Licata, Herbario Universitario PORT (Guanare, Edo. Portuguesa)
BS – Basil Stergios, Herbario Universitario PORT (Guanare, Edo. Portuguesa)
NC – Nancy Chacón, Herbario Regional de Guayana (GUYN)
GA – Gerardo Aymard, Herbario Universitario PORT (Guanare, Edo. Portuguesa)

Specimen Condition or other evidence
Fe – fertile (ferns)
Fl – flower
Fr – fruit
St – sterile
O – observation/(specimen not collected)

Preliminary list of plants observed or collected
during the RAP expedition to the Eastern Kanuku
Mountains, Guyana with site and habitat information

	Collection Locality Habitats		Specimen number/condition	Determiner	New Record for Kanukus
	Pobowau Creek	Cacique Mountain			
Acanthaceae					
Ruellia sp.	SFM		5456wd/Fl	WD	
Anacardiaceae					
Tapirira guianensis Aubl.	SV		O	WD	
Annonaceae					
Anaxagorea cf. *brevipes* Benth.	SFM	SFM	5638wd/Fr	WD	
Annona hypoglauca Mart.	RM		5530wd/Fl	WD	
Duguetia calycina Benoist	SFM		5461wd/Fl	WD	
Duguetia cauliflora R.E. Fr.		MF	5561wd/Fl	WD	
Duguetia pycnastera Sandwith		SFM	5639wd/Fl	ES	X
Duguetia sp.	SFM	MFS, MFCM, SFM	5615wd/Fl	WD	
Unonopsis sp.		MF	5558/Fr	WD	
Apocynaceae					
Aspidosperma sp. 1	SFM	MFS, MFCM, SFM	O	WD	
Aspidosperma sp. 2	SFM		O		
Mesechites trifida (Jacq.) Müll. Arg.	PC		5539wd/Fl	WD	
Tabernaemontana attenuata (Miers) Urb.		SFM	5635wd/Fl	WD	X
Tabernaemontana sp.	RM	MFCM	5519wd/St	WD	
Unknown Apocynaceae sp. 1		MFS	O	WD	
Unknown Apocynaceae sp. 2		MFS	5619wd/St	WD	
Unknown Apocynaceae sp. 3		SFM	O		
Araceae					
Heteropsis flexuosa (Kunth) G.S. Bunting	SFM		5511wd/Fr	WD	
Montrichardia arborescens (L) Schott.	PC		5550wd/Fr	WD	X
Philodendron sp.		SV	5588wd/Fl	WD	
Araliaceae					
Schefflera morototoni		MFS	O	WD	
Arecaceae					
Bactris sp.	SFM		O	WD	
Geonoma baculifera (Poit.) Kunth.	SFM		5474wd/Fl	WD	
Geonoma sp.	SFM	MFS, MFCM, SFM	O	WD	
Iriartella sp.	SFM		O	WD	
Bignoniaceae					
Clytostoma binatum (Thumb.) Sandw.	R		5460wd/Fl	WD	
Cydista aequinoctialis (L.) Miers.	RM, PC		5513wd/Fl	WD	
Jacaranda copaia		SV	O	WD	
Bixaceae					
Bixa orellana Willd.	RM		5522wd/Fl	WD	X
Bombacaceae					
Catostemma fragrans Benth.	SFM	SFM, MFS, MFCM	O	WD	X

continued

	Collection Locality Habitats		Specimen number/condition	Determiner	New Record for Kanukus
	Pobowau Creek	Cacique Mountain			
Boraginaceae					
Cordia sp.	RM	MFS	O		
Tournefortia bicolor Sw.	SFM		5483wd/Fl	WD	X
Tournefortia sp.	PC		5551wd/Fl	WD	
Bromeliaceae					
Aechmea bromeliifolia (Rudge) Baker		SV	5597wd/Fr	WD	
Bromelia sp.		SV	O	WD	
Guzmania sp.		SV	5582wd/Fl	WD	
Pitcairnia bulbosa L.B. Sm.		SV	5581wd/Fl	WD	
Burseraceae					
Protium heptaphyllum	SFM		O	WD	
Trattinickia sp.		MFS, SFM	O	WD	
Cactaceae					
Epiphyllum phyllanthus (L.) Haw.		SV	O	WD	X
Hylocereus scandens (Salm-Dyck) Backeb		SV	O	WD	X
Caesalpiniaceae					
Brownea sp.		R	5623wd/Fl	WD	X
Dialium guianense (Aubl.) Sandwith	PC		5548wd/Fr.	WD	
Heterostemon mimosoides Desf.	SFM		5479wd/St	WD	X
Macrolobium acaciifolium (Benth.) Benth.	PC		5543wd/Fl	WD	
Mora excelsa Benth.	SFM		O	WD	
Tachigali sp.		SFM	O	WD	
Caryocaraceae					
Caryocar sp.		MFS	O	WD	
Cecropiaceae					
Cecropia peltata L.	SFM, RM		O	WD	
Celastraceae					
Maytenus guyanensis Klotzsch ex Reissek	SFM		5512wd/Fl	WD	
Chrysobalanaceae					
Hirtella racemosa Lam.		MF	5557wd/Fl	WD	
Hirtella sp.	SFM		O	WD	
Licania cf. *lasseri* Maguire	PC		5538wd/Fl	WD	X
Licania sp. 1	SFM, PC	R, MFS	O	WD	
Licania sp. 2	MFS, SFM	MFCM, MFS	O	WD	
Clusiaceae					
Clusia sp.		SV	5580wd/Fl	WD	
Rheedia sp.		MFCM, SFM	O	WD	
Vismia sp.	SFM		O	WD	
Combretaceae					
Combretum laxum Jacq.	SFM	R	5506wd/Fr	WD	
Combretum sp.	PC		5545wd/Fr	WD	

continued

Preliminary list of plants observed or collected
during the RAP expedition to the Eastern Kanuku
Mountains, Guyana with site and habitat information

	Collection Locality Habitats		Specimen number/condition	Determiner	New Record for Kanukus
	Pobowau Creek	Cacique Mountain			
Terminalia sp.		SV	O	WD	
Unknown Combretaceae sp. 1	SFM		5477awd/St	WD	
Unknown Combretaceae sp. 2		MFS	O	WD	
Costaceae					
Costus spiralis (Jacq.) Roscoe var. *spiralis*	SFM, RM		5528wd/Fl	WD	X
Cyperaceae					
Fuirena sp.	RM		5518wd/Fr	WD	
Scleria sp. 1	RM	SV	5527wd/Fr	WD	
Scleria sp. 2		MFCM	O	WD	
Dichapetalacea					
Tapura guianensis Aubl.	SFM		5461wd/Fl	AL, WD	
Elaeocarpaceae					
Sloanea sp.		MFCM, SFM	O	WD	
Erythroxylaceae					
Erythroxylum sp. 1		SV	O	WD	
Erythroxylum sp. 2		SV	O	WD	
Euphorbiaceae					
Dalechampia scandens L.	RM		5526wd/Fl	WD	
Mabea montana Müll. Arg.	PC	R	5547wd/Fr	WD	
Sagotia racemosa Baill.	SFM	SFM	5510wd/Fr	WD	
Sagotia cf. *racemosa*		MFCM, MFS, SFM	O	WD	
Fabaceae					
Aldina sp.	SFM		5491wd/St	BS	
Clathrotropis brachypetala (Tul.) Kleinhoonte	SFM	MFCM	5498wd/St	WD	
Clathrotropis macrocarpa Ducke		MFCM, SFM, MFS	5612wd/St	WD	X
Dioclea guianensis Aubl.	RM, PC, SFM		5482wd/Fl	WD	
Lonchocarpus sp.	SFM	R	5645wd/St	WD	
Machaerium inundatum (Mart. ex Benth.) Ducke		R	5573wd/Fr	WD	
Pterocarpus rohrii Vahl		SFM	5637wd/St	WD	
Swartzia sp.	SFM		O		
Unknown Fabaceae sp. 1		MFS	C	WD	
Unknown Fabaceae sp. 2		SFM	O	WD	
Unknown Fabaceae sp. 3		SFM	O	WD	
Unknown Fabaceae sp. 4		SFM	O	WD	
Unknown Fabaceae sp. 5		SFM	5621wd/Fl	WD	
Flacourtiaceae					
Homalium racemosum Jacq.	PC		5552wd/Fl	WD	X
Laetia sp.		SFM	5628wd/St	WD	X
Gesneriaceae					
Drymonia coccinea (Aubl.) Wiehler		R	5575wd/Fl	WD	X

continued

	Collection Locality Habitats		Specimen number/condition	Determiner	New Record for Kanukus
	Pobowau Creek	Cacique Mountain			
Hippocrateaceae					
Cuervea kappleriana (Miq.) A.C. Smith	R		5507wd/Fl	WD	X
Cuervea sp.	SFM		5537wd/Fl	WD	X
Pristimera nervosa (Miers) A.C. Smith	PC		5486wd/Fl	WD	X
Lacistemaceae					
Lacistema sp.	SFM		5464wd/St	WD	
Lauraceae					
Aniba cf. *guianensis*		MFCM	5603wd/Fl	WD	
Licaria polyphylla (Nees) Kosterm.	R	MFS	5570wd/Fl	WD	X
Ocotea sp.	SFM		O	WD	
Unknown Lauraceae sp. 1	SFM		5471wd/St	WD	
Unknown Lauraceae sp. 2	SFM		5501wd/St	WD	
Unknown Lauraceae sp. 3		MFCM	O	WD	
Unknown Lauraceae sp. 4		MFS	O	WD	
Unknown Lauraceae sp. 5		SFM	O	WD	
Lecythidaceae					
Bertholletia excelsa Bonpl.	SFM		O	WD	
Eschweilera sp. 1	SFM		5465wd/St	WD	
Eschweilera sp. 2		MFS, MFCM, SFM	5458wd/Fl	WD	
Lindsaeaceae					
Lindsaea sp.		MFS	O	WD	
Loganiaceae					
Strychnos sp.	SFM		5485wd/St	WD	X
Lycopodiaceae					
Huperzia cf. *dichotoma* (Jacq.) Trevis.	SFM		5452wd/St	WD	X
Malpighiaceae					
Banisteriopsis martiniana (A. Juss.) Cuatr.		R	5564wd/Fl	WD	
Malvaceae					
Hibiscus bifurcatus Cav.	RM		5520wd/Fl	WD	
Marantaceae					
Calathea variegata		MFCM	O	WD	
Calathea sp.	SFM		O	WD	
Ischnosiphon arouma	SFM	MFS, MFCM, SFM	O	WD	
Unknown Marantaceae sp. 1		SFM	O	WD	
Unknown Marantaceae sp. 2		SFM	O	WD	
Melastomataceae					
Miconia eugenioides? Triana	SFM		5525wd/Fr	WD	X
Mouriri sp.	RM		5480wd/Fl	WD	
Ossaea micrantha? (Sw.) Macfad.	SFM		5483wd/Fr	WD	X

continued

Preliminary list of plants observed or collected
during the RAP expedition to the Eastern Kanuku
Mountains, Guyana with site and habitat information

	Collection Locality Habitats				
	Pobowau Creek	Cacique Mountain	Specimen number/condition	Determiner	New Record for Kanukus
Meliaceae					
Carapa guianensis Aubl.	SFM	MFS, MFCM, SFM	O	WD	
Guarea guidonia (L.) Sleumer	SFM, SFM	SFM, MF, MFS, MFCM,	5633wd/Fr	WD	
Guarea kunthiana A. Juss.	SFM		5556wd/Fr	WD	
Guarea sp.		MFS	5472wd/St	WD	
Trichilia schomburgkii subsp. *schomburgkii*	SFM	MF	5560wd/Fl	WD	
Trichilia sp.		MFCM	5457wd/Fl	WD	
Unknown Meliaceae sp. 1		SFM	O	WD	
Unknown Meliaceae sp. 2		SFM	O	WD	
Unknown Meliaceae sp. 3		SFM	O	WD	
Mimosaceae					
Entada polyphylla Desf.	RM		5524wd/Fl	WD	
Inga ingoides (Rich.) Willd.	PC		5546wd/Fl	WD	
Inga laurina (Sw.) Willd.	PC		5533wd/Fl	WD	X
Inga sp. 1	SFM, RM		5500wd/St	WD	
Inga sp. 2		MFCM, MFS, SFM	O	WD	
Parkia sp.	SFM		5477wd/St	WD	
Zapoteca formosa? (Kunth) H.M. Hern.			5541wd/Fl	WD	X
Zygia cf. *unifoliolata* (Benth.) Pittier		SV	5587wd/Fl	WD	
Monimiaceae					
Siparuna sp.		MFS	5617wd/Fl	WD	
Moraceae					
Ficus sp. 1		MFS	O	WD	
Ficus sp. 2		SFM	5618wd/St	WD	
Maquira sp.		MFCM	5600wd/St	WD	
Myristicaceae					
Iryanthera sp.	SFM		O	WD	
Virola sp.		MFS, MFCM, SFM	5620wd/St	WD	
Myrsinaceae					
Stylogine longifolia	SFM		5489wd/Fr	WD	
Myrtaceae					
Calyptranthes multiflora? O. Berg.	PC		5542wd/Fl	WD	X
Eugenia sp.		SV	5577wd/Fl	WD	
Myrcia sp.	SFM		5475wd/St	WD	
Psidium persoonii McVaugh	PC		5536wd	WD	X
Psidium sp.	SFM, PC	SFM	5502wd/St	WD	
Unknown Myrtaceae sp.		R	5622wd/Fl	WD	
Olacaceae					
Heisteria acuminata? Benth. & Hook.f		SFM	5643wd/Fr	WD	X

continued

	Collection Locality Habitats		Specimen number/condition	Determiner	New Record for Kanukus
	Pobowau Creek	Cacique Mountain			
Orchidaceae					
Prosthechea vespa (Vell.) Higgins		SV	5593wd/Fl	NC	X
Pleurothallis sp.		SV	5595wd/Fl	WD	
Unknown Orchidaceae sp. 1	SFM		5454wd/Fl	WD	
Unknown Orchidaceae sp. 2	SFM		5455wd/Fl	WD	
Unknown Orchidaceae sp. 3		SV	O	WD	
Unknown Orchidaceae sp. 4		SV	5589wd/Fl	WD	
Unknown Orchidaceae sp. 5		SV	5596wd/Fl	WD	
Passifloraceae					
Passiflora bomareifolia Steyerm. & Maguire	RM	MFCM	5598wd/Fl	WD	X
Piperaceae					
Piper hispidum Sw.	RM		5523wd/Fl	WD	
Piper hostmannianum (Miq.) C. DC.	RM		5529wd/Fl	WD	X
Poaceae					
Ichnanthus nemoralis (Schrad) Hitchc. & Chase	RM	SV	5517wd/Fl	WD	
Neurolepis angustifolia Swallen	SFM		O	WD	X
Olyra ciliatifolia (Raddi)		SV	5583wd/Fr	WD	
Olyra sp.		MF, MFCM, MFS	C, O	WD	
Panicum pilosum Sw.	RM		5516wd/Fl	WD	
Podostemaceae					
Apinagia sp. *			5644wd/St	WD	X
Polygonaceae					
Coccoloba sp.	RM	SV	5571wd/Fr	WD	
Triplaris cf. *weigeltiana* (Rchb.) Kuntze	RM	SV	5521wd/St	WD	
Polypodiaceae					
Microgramma lycopodioides (L.) Copel.	SFM		5453wd/Fe	WD	
Microgramma reptans (Cav.) A.R. Sm.		MF	5562wd/Fe	WD	
Pecluma pectinata (L.) M.G. Price		SFM	5640wd/Fe	WD	
Pleopeltis cf. *percussa* (Cav.) Hook & Grev.	SFM		5451wd/Fe	WD	
Polypodium sp.		SV	5586wd/Fe	WD	
Unknown Polypodiaceae sp.		SV	5594wd/Fe	WD	
Pteridaceae					
Adiantum sp.	SFM	MFCM, MFS, SFM	5610wd/Fe	WD	
Rhizoporaceae					
Cassipourea guianensis Aubl.	SFM		5459wd/Fl	AL	
Rubiaceae					
Alibertia latifolia (Benth.) K. Schum.	SFM, PC		5488wd/Fl	WD	X
Diodia sp.	RM		O	WD	
Faramea capillipes Müll. Arg.	SFM		5504wd/Fl	WD	
Faramea occidentalis (L.) A. Rich.	PC		5535wd/Fl	WD	X
Genipa spruceana Steyerm.	PC		5534wd/Fl	WD	
Geophyla sp.		MFS	5646wd/Fr	WD	

continued

Preliminary list of plants observed or collected
during the RAP expedition to the Eastern Kanuku
Mountains, Guyana with site and habitat information

| | Collection Locality Habitats | | | | |
	Pobowau Creek	Cacique Mountain	Specimen number/condition	Determiner	New Record for Kanukus
Morinda tenuiflora (Benth.) Steyerm.		SV, MFS	5647wd/Fr	WD	
Palicourea cf. *crocea* (Sw.) Roem & Schult.	PC		5544wd/Fr	WD	X
Psychotria apoda Steyerm.		R	5576wd/Fl	WD	
Psychotria lindenii? Standl.	SFM	SV	5584wd/Fl	WD	X
Psychotria sp.		SFM	5639wd/Fl	WD	
Unknown Rubiaceae sp.		MFCM	5609wd/Fl	WD	
Sapindaceae					
Matayba adenanthera Radlk.	PC		5549wd/Fr	WD	X
Matayba opaca Radlk.	R		5509wd/Fl	WD	
Toulicia pulvinata Radlk.	PC	R	5531wd/Fr	WD	
Unknown Sapindaceae sp. 1		SFM	O	WD	
Unknown Sapindaceae sp. 2	R		5572wd/Fr	WD	
Sapotaceae					
Pouteria venosa (Mart.) Baheni subsp. *amazonica*		SFM	5626wd/Fl	GA	
Pouteria sp. 1		MFS	O	WD	
Pouteria sp. 2	SFM		5503wd/St	WD	
Pouteria sp. 3		MFCM	5601wd/St	WD	
Unknown Sapotaceae sp. 1		MFCM	5605wd/St	WD	
Unknown Sapotaceae sp. 2		MFS, SFM	O	WD	
Schizaeaceae					
Anemia ferruginea		SV	5591wd/Fe	WD	
Selaginellaceae					
Selaginella sp.	SFM	MFS, MFCM, SFM	5607wd/Fe	WD	
Solanaceae					
Solanum sp.	RM		O	WD	
Sterculiaceae					
Byttneria divaricata Benth.	RM	R	5563wd/Fr	WD	
Sterculia sp.		MFCM, MFS, SFM	O	WD	
Strelitziaceae					
Phenakospermun guyannense (Rich.) Endl. ex Miq.		MFS	O	WD	
Thelypteridaceae					
Thelypteris sp.		MFCM	5611wd/Fe	WD	
Theophrastaceae					
Clavija sp.	SFM	SFM	O	WD	
Violaceae					
Payparola sp.		SFM	5629wd/Fr	WD	
Rinorea flavescens (Aubl.) Kuntze		MFCM	5602wd/Fl	WD	X
Rinorea lindeniana (Tul.) Kuntze		SFM	5631wd/Fl	WD	
Rinorea riana (DC. ex Ging.) Kuntze, nom. illeg.	SFM	R, SFM	5642wd/Fl	WD	
Rinorea sp.	SFM	MFS, SFM	O	WD	

continued

	Collection Locality Habitats		Specimen number/condition	Determiner	New Record for Kanukus
	Pobowau Creek	Cacique Mountain			
Vitaceae					
Cissus erosa Rich.	R, RM		5508wd/Fl	WD	
Vittariaceae					
Vittaria sp.	SFM	SV	5450wd/Fe	WD	
Zingiberaceae					
Renealmia floribunda K. Schum.		SFM	5627wd/Fr	WD	

Appendix 3

Fishes of the Lower Kwitaro and Rewa Rivers per collection site (S1–S8), Eastern Kanuku Mountains, Guyana

Jan H. Mol

Values indicate numbers of individuals collected; p = presence confirmed

(S1)   Pobawau Creek (3º16'3.1"N, 58º46'42.7"W)
(S2)   a bay in the Kwitaro River 500 m upstream of Pobawau Creek
(S3)   a shallow, short-cut connection between two bends of the Kwitaro River
(S4)   the Kwitaro River at Cacique Mountain camp (3º11'29.5"N, 58º48'42.0"W)
(S5)   a lake connected to the Kwitaro River (approximately 700 m upstream of S4)
(S6)   a small, unnamed forest creek (elevation 110 m) at the base of the Cacique Mountain slope
(S7)   the Rewa River at the Corona Falls rapid (3º11'35.0"N, 58º48'39.6"W)
(S8)   an unnamed forest creek at the Corona Falls rapid (same coordinates as S7)

Taxon	S1	S2	S3	S4	S5	S6	S7	S8
Characiformes								
Anostomidae								
Anostomus ternetzi	-	-	-	-	1	-	-	-
Leporinus friderici	3	-	1	-	1	-	-	-
Leporinus cf. *granti*	-	-	-	-	-	p	p	p
Leporinus pellegrini	-	-	1	-	-	-	-	-
Schizodon fasciatum	5	-	-	-	-	-	-	-
Characidae								
Acestrorhynchus guianensis	3	-	-	-	4	-	-	-
Aphyocharax erythrurus	-	-	-	2	1	-	-	p
Aphyodite grammica	-	-	10	-	-	-	-	-
Astyanax bimaculatus	1	-	-	-	-	34	-	2
Astyanax guianensis	3	-	-	-	-	-	-	-
Astyanax cf. *mutator*	-	-	1	-	-	-	-	-
Astyanax polylepis	7	-	1	-	-	-	-	3
Astyanax sp.	2	-	-	-	-	-	-	-
Boulengerella lucia	1	-	-	2	-	-	p	-
Brachychalcinus guianensis	1	-	-	-	-	-	-	-
Brycon falcatus	10+	-	-	2	-	-	p	-
Bryconops affinis	2	-	1	-	-	-	-	-
Bryconops caudomaculatus	-	-	-	-	-	-	-	14

continued

Taxon	S1	S2	S3	S4	S5	S6	S7	S8
Chalceus macrolepidotus	-	-	-	-	2	-	-	-
Charax gibbosus	1	-	-	-	2	-	-	-
Cynodon gibbus	8	1	-	-	10	-	-	-
Hemigrammus bellottii	27	-	-	-	-	3	-	-
Hemigrammus cf. *boesemani*	-	-	13	-	-	-	-	-
Hemigrammus guyanensis	1	-	-	-	-	-	-	-
Hemigrammus rodwayi	1	-	-	-	-	1	-	-
Hyphessobrycon serpae	39	-	-	-	-	-	-	-
Hyphessobrycon simulatus	-	-	1	-	-	-	-	-
Hyphessobrycon cf. *tenuis*	8	-	-	-	-	-	-	-
Microschemobrycon geisleri	-	-	15	-	-	-	-	-
Moenkhausia collettii	4	-	-	1	-	-	-	5
Moenkhausia cf. *copei*	-	-	36	-	-	-	-	-
Moenkhausia cf. *cotinho*	-	-	-	-	-	-	-	4
Moenkhausia georgiae	1	-	-	-	-	-	-	-
Moenkhausia grandisquamis	13	-	3	-	1	-	-	11
Moenkhausia hemigrammoides	-	-	-	-	-	3	-	-
Moenkhausia intermedia	1	-	38	-	-	-	-	-
Moenkhausia lepidura	35	-	-	-	60	-	-	-
Moenkhausia oligolepis	-	-	-	-	-	5+	-	3
Moenkhausia aff. *surinamensis*	1	-	-	-	-	-	-	-
Phenacogaster microstictus	-	-	2	1	-	-	-	-
Poptella brevispinnis	4	-	1	-	3	-	-	-
Roestes ogilviei	-	-	-	-	4	-	-	-
Tetragonopterus chalceus	3	-	2	-	-	-	-	-
Triportheus rotundatus	2	-	-	2	10+	-	-	-
unidentified characin	1	-	-	-	-	-	-	-
Characidiidae								
Characidium fasciadorsale	-	-	-	1	-	-	-	-
Characidium pellucidum	1	-	-	-	-	-	-	-
Crenuchidae								
Crenuchus spilurus	-	-	-	-	-	19	-	-
Curimatidae								
Curimata cilliata	10	-	-	-	30+	-	-	-
Curimata cyprinoides	2	-	-	-	2	-	-	-
Cyphocharax microcephala	-	-	-	-	-	-	-	1
Cyphocharax spilurus	-	-	6	-	-	-	-	2
Prochilodus rubrotaeniatus	-	-	-	-	7	-	-	-
Erythrinidae								
Erythrinus erythrinus	-	-	-	-	-	1	-	-
Hoplerythrinus unitaeniatus	-	-	-	-	-	2	-	-
Hoplias aimara	-	-	-	p	-	-	1	-
Hoplias malabaricus	-	-	-	p	-	2	p	-

continued

Taxon	S1	S2	S3	S4	S5	S6	S7	S8
Gasteropelecidae								
Carnegiella strigata	1	-	-	1	p	5+	-	18
Hemiodidae								
Hemiodopsis quadrimaculatus	-	-	-	-	-	-	p	-
Hemiodopsis sp. (cf. *gracilis*)	-	-	-	-	-	-	-	p
Parodon guyanensis	-	-	-	-	-	-	-	p
Lebiasinidae								
Nannostomus marginatus	-	-	-	-	-	5+	-	-
Pyrrhulina stoli	-	-	-	-	-	14	-	-
Serrasalmidae								
Colossoma bidens	-	-	-	-	-	-	p	-
Myleus rubripinnis	2	1	-	-	-	-	-	-
Serrasalmus humeralis	10+	-	-	-	2	-	-	-
Serrasalmus rhombeus	10+	p	-	p	p	-	1	-
Gymnotiformes								
Electrophoridae								
Electrophorus electricus	-	-	-	p	-	-	p	10+
Gymnotidae								
Gymnotus carapo	-	-	-	-	-	3	-	-
Rhamphichthyidae ?								
Gymnorhampichthys hypostomus	-	-	-	-	-	-	-	2
Sternopygidae								
Eigenmannia sp.	-	-	-	-	-	-	-	1
Siluriformes								
Ageneiosidae								
Ageneiosus inermis	2	-	-	p	3	-	1	-
Auchenipteridae								
Auchenipterus nuchalis	-	-	-	-	1	-	-	-
Parauchenipterus galeatus	2	-	-	-	-	-	-	-
Callichthyidae								
Callichthys callichthys	-	-	-	p	-	-	-	-
Corydoras bondi bondi	-	-	7	-	-	-	-	-
Corydoras melanistius	3	-	1	-	-	-	-	p
Megalechis sp.	-	-	-	p	-	-	-	-
Hypophthalmidae								
Hypophthalmus edentatus	-	-	-	-	2	-	-	-
Loricariidae								
Ancistrus hoplogenys	-	-	-	-	-	-	-	3
Ancistrus sp.	1	-	-	-	-	-	-	-
Hypoptopoma guianense	10+	-	-	-	35	-	-	-
Hypostomus gymnorhynchus	-	-	-	-	-	-	-	1
Hypostomus cf. *hemiurus*	2	-	-	-	-	-	-	-
Hypostomus cf. *micropunctatus*	-	1	-	-	1	-	-	-

continued

Taxon	S1	S2	S3	S4	S5	S6	S7	S8
Hypostomus cf. *plecostomus*	1	-	-	-	-	-	-	-
Hypostomus ventromaculatus	-	-	-	-	-	-	-	1
Pimelodidae								
Hemisorubim platyrhynchos	-	-	-	p	-	-	-	-
Paulicea sp.	-	-	-	1	-	-	-	-
Pimelodella cristata	1	-	6	-	-	-	-	p
Pimelodella geryi	-	-	1	-	-	-	-	-
Pimelodella macturki	-	-	2	-	-	-	-	-
Pimelodus albofasciatus	1	-	7	-	2	-	-	-
Practocephalus hemiliopterus	-	1	-	-	-	-	-	-
Pseudoplatystoma fasciatum	-	-	-	3	-	-	p	1
Pseudoplatystoma tigrinum	-	-	-	p	-	-	p	-
Rhamdia quelen	-	-	-	-	-	1	-	-
Sorubim lima	5	-	-	-	1	-	p	-
Trichomycteridae								
Vandellia cf. *plazaii*	8	-	-	-	-	-	-	-
Perciformes								
Cichlidae								
Aequidens tetramerus	-	-	-	-	-	2	-	-
Apistogramma cf. *ortmanni*	-	-	-	-	-	-	-	3
Apistogramma steindachneri	3	-	-	-	-	4	-	-
Cichla ocellaris	-	-	-	-	-	-	1	-
Cichla temensis	-	-	-	-	-	-	p	-
Crenicichla alta	-	-	-	-	-	1	-	-
Crenicichla wallacei	-	-	-	-	-	-	-	3
Sciaenidae								
Plagioscion surinamensis	-	-	-	-	12	-	-	-
Miscellaneous Groups								
Belonidae								
Potamorrhaphis guianensis	2	-	-	-	-	-	-	2
Dasyatidae								
Potamotrygon histrix	-	-	-	1	-	-	p	-
Potamotrygon cf. *motoro*	-	-	-	p	-	-	p	-
Osteoglossidae								
Arapaima gigas	-	-	-	-	p	-	-	-
Osteoglossum bicirrhosum	-	-	-	p	p	-	-	-
Tetraodontidae								
Colomesus asellus	-	-	-	-	1	-	-	-

Appendix 4

Bird species recorded for the Eastern Kanuku Mountains, Lower Kwitaro River, Guyana

Davis W. Finch, Wiltshire Hinds, Jim Sanderson, and Olivier Missa

The abundance of each species is indicated by a letter code (see below) for each site surveyed during the 20–29 September 2001 RAP expedition: Pobawau Creek, Cacique Mountain Forest, and Cacique Lowland Forest. The species recorded during two other surveys conducted by Davis Finch in November 1998 ('98) and 2001 ('01) are also listed. The habitats where the species are usually found are indicated and follow Braun et al. (2000). Common and scientific names and order follow Clements (2000).

* Restricted to the Guianas and adjacent Venezuela and Brazil
** Restricted to north of the Amazon

Habitats

LF	Lowland forest, including both terra firme and seasonally flooded forest
MF	Montane forest
RI	Riverine habitats, including waters, islands, banks, waterfalls, and riparian forests
MA	Marine or salt water habitats, including coastal and pelagic waters
MU	Mudflats and coastal beaches
FW	Fresh water habitats, including lakes, conservancies, ponds, oxbows, marshes, and canals
MN	Mangrove forest
HU	Habitats altered by humans, such as gardens, towns, roadsides, agricultural lands, disturbed forests, and forest edge
SV	Savannah grasslands
SC	Scrub or brush habitats, including white sand scrub, bush islands, and dense, low second growth
PA	Palm trees and forests

Abundance Codes

C	Common, more than 10 individuals encountered daily
F	Fairly common, fewer than 10 individuals encountered daily
U	Uncommon, small number of individuals encountered (not daily)
S	Scarce, 1–3 individuals encountered at the site
O	Indicates species encountered only once during the entire survey

Common name	Scientific name	Finch's surveys	Pobawau Creek	Cacique Mountain Forest	Cacique Lowland Forest	Habitats
Tinamous	**Tinamidae (4)**					
Great Tinamou	*Tinamus major*	'01	U	U	U	LF
Cinereous Tinamou	*Crypturellus cinereus*	'01	U	-	-	LF SC
Variegated Tinamou	*Crypturellus variegatus*	'98 '01	U	-	U	SC LF
Red-legged Tinamou	*Crypturellus erythropus*	-	U	-	U	LF SC
Cormorants	**Phalacrocoracidae (1)**					
Neotropic Cormorant	*Phalacrocorax brasilianus*	-	F	-	U	FW RI
Anhingas	**Anhingidae (1)**					
Anhinga	*Anhinga anhinga*	'98 '01	F	-	F	FW RI
Herons	**Ardeidae (7)**					
Rufescent Tiger-Heron	*Tigrisoma lineatum*	'01	O	-	-	RI
Cocoi Heron	*Ardea cocoi*	'98 '01	F	-	U	FW RI
Snowy Egret	*Egretta thula*	-	U	-	-	MU MN FW
Striated Heron	*Butorides striatus*	'98 '01	F	-	U	FW MN RI
Agami Heron	*Agamia agami*	-	-	-	O	LF RI
Capped Heron	*Pilherodius pileatus*	'98 '01	U	-	U	FW RI
Boat-billed Heron	*Cochlearius cochlearius*	'01	-	-	-	FW MN
Ibises	**Threskiornithidae (1)**					
Green Ibis	*Mesembrinibis cayennensis*	'98 '01	F	-	U	LF RI
Storks	**Ciconiidae (1)**					
Jabiru	*Jabiru mycteria*	-	O	-	-	FW
Ducks, Geese	**Anatidae (1)**					
Muscovy Duck	*Cairina moschata*	-	O	-	O	FW RI
Vultures	**Cathartidae (4)**					
Black Vulture	*Coragyps atratus*	'98 '01	C	U	U	SC HU
Turkey Vulture	*Cathartes aura*	-	U	U	U	HU SC SV
Greater Yellow-headed Vulture	*Cathartes melambrotus*	'98 '01	U	U	U	LF
King Vulture	*Sarcoramphus papa*	'98 '01	U	U	U	LF SV
Ospreys	**Pandionidae (1)**					
Osprey	*Pandion haliaetus*	'98 '01	U	-	-	MA FW RI
Hawks, Eagles	**Accipitridae (8)**					
Gray-headed Kite	*Leptodon cayanensis*	'98	-	-	-	LF
Double-toothed Kite	*Harpagus bidentatus*	'98	-	-	-	LF MF
Great Black-Hawk	*Buteogallus urubitinga*	'98 '01	-	-	O	LF RI
Gray Hawk	*Asturina nitida*	-	O	-	-	SC HU
Roadside Hawk	*Buteo magnirostris*	'98 '01	O	-	-	HU SC
Harpy Eagle	*Harpia harpyja*	-	-	-	O	LF
Ornate Hawk-Eagle	*Spizaetus ornatus*	-	-	O	-	LF MF
Black-and-white Hawk-Eagle	*Spizastur melanoleucus*	'98	-	-	-	LF MF
Falcons, Caracaras	**Falconidae (8)**					
Barred Forest-Falcon	*Micrastur ruficollis*	'98	-	-	-	LF

continued

Common name	Scientific name	Finch's surveys	Pobawau Creek	Cacique Mountain Forest	Cacique Lowland Forest	Habitats
Lined Forest-Falcon	*Micrastur gilvicollis*	'98	-	-	O	LF
Collared Forest-Falcon	*Micrastur semitorquatus*	'98	-	-	-	LF SC
Black Caracara	*Daptrius ater*	'98 '01	U	-	U	RI LF SC
Red-throated Caracara	*Daptrius americanus*	'98 '01	-	-	U	LF
Laughing Falcon	*Herpetotheres cachinnans*	-	U	-	-	LF SC
Bat Falcon	*Falco rufigularis*	-	U	-	-	LF SC RI
Orange-breasted Falcon	*Falco deiroleucus*	-	U	-	-	LF MF
Curassows, Guans	**Cracidae (4)**					
Little Chachalaca	*Ortalis motmot*	'98 '01	U	-	F	LF SC
Spix's Guan	*Penelope jacquacu*	-	U	U	U	LF
Blue-throated Piping-Guan	*Pipile cumanensis*	-	O	-	-	LF MF
Black Curassow**	*Crax alector*	-	U	-	U	LF MF
Trumpeters	**Psophiidae (1)**					
Gray-winged Trumpeter	*Psophia crepitans*	'98	-	-	U	LF
Sungrebes	**Heliornithidae (1)**					
Sungrebe	*Heliornis fulica*	-	U	-	S	FW RI
Sunbitterns	**Eurypygidae (1)**					
Sunbittern	*Eurypyga helias*	'98 '01	U	-	U	RI LF
Plovers	**Charadriidae (1)**					
Pied Lapwing	*Vanellus cayanus*	'98 '01	U	-	O	RI
Sandpipers	**Scolopacidae (2)**					
Solitary Sandpiper	*Tringa solitaria*	-	F	-	U	FW RI
Spotted Sandpiper	*Actitis macularia*	'98 '01	F	-	U	RI MN FW
Terns	**Sternidae (2)**					
Yellow-billed Tern	*Sterna superciliaris*	-	U	-	U	RI FW
Large-billed Tern	*Phaetusa simplex*	'01	U	-	-	RI FW
Skimmers	**Rhynchopidae (1)**					
Black Skimmer	*Rynchops niger*	-	U	-	-	MA MU RI
Pigeons, Doves	**Columbidae (4)**					
Scaled Pigeon	*Columba speciosa*	'01	-	-	-	LF SC SV
Plumbeous Pigeon	*Columba plumbea*	'98 '01	U	-	U	LF MF
Ruddy Pigeon	*Columba subvinacea*	'01	F	-	F	LF MF
Gray-fronted Dove	*Leptotila rufaxilla*	'01	U	-	U	LF
Parrots	**Psittacidae (12)**					
Painted Parakeet	*Pyrrhura picta*	'98 '01	U	-	U	LF MF
Red-and-green Macaw	*Ara chloroptera*	'98 '01	F	U	F	LF
Scarlet Macaw	*Ara macao*	'98 '01	F	-	U	LF
Blue-and-yellow Macaw	*Ara ararauna*	'98	U	-	U	PA LF RI
Golden-winged Parakeet	*Brotogeris chrysopterus*	'98 '01	F	-	F	LF
Black-headed Parrot**	*Pionites melanocephala*	'98 '01	U	-	U	LF
Caica Parrot*	*Pionopsitta caica*	'98 '01	U	-	U	LF
Blue-headed Parrot	*Pionus menstruus*	'98 '01	U	-	U	LF

continued

Common name	Scientific name	Finch's surveys	Pobawau Creek	Cacique Mountain Forest	Cacique Lowland Forest	Habitats
Dusky Parrot	*Pionus fuscus*	'98	U	-	O	LF
Orange-winged Parrot	*Amazona amazonica*	'01	F	-	F	LF SC
Mealy Parrot	*Amazona farinosa*	'98	F	-	F	LF
Red-fan Parrot	*Deroptyus accipitrinus*	'98	U	-	U	LF
Cuckoos	**Cuculidae (4)**					
Squirrel Cuckoo	*Piaya cayana*	'98 '01	S	-	S	LF
Black-bellied Cuckoo	*Piaya melanogaster*	'98	-	-	O	LF
Little Cuckoo	*Piaya minuta*	'01	-	-	-	SC FW
Greater Ani	*Crotophaga major*	'01	F	-	U	RI MN FW
Typical Owls	**Strigidae (5)**					
Tropical Screech-Owl	*Otus choliba*	-	S	-	-	SC LF
Tawny-bellied Screech-Owl	*Otus watsonii*	'98	-	-	-	LF
Crested Owl	*Lophostrix cristata*	'01	-	-	-	LF MF
Spectacled Owl	*Pulsatrix perspicillata*	'98 '01	S	-	O	LF
Amazonian Pygmy-Owl	*Glaucidium hardyi*	'01	-	-	-	LF MF
Potoos	**Nyctibiidae (1)**					
Great Potoo	*Nyctibius grandis*	-	-	-	O	LF
Nighthawks, Nightjars	**Caprimulgidae (4)**					
Short-tailed Nighthawk	*Lurocalis semitorquatus*	'01	U	-	-	LF RI
Common Pauraque	*Nyctidromus albicollis*	'01	U	-	U	SC HU
Blackish Nightjar	*Caprimulgus nigrescens*	-	U	-	U	LF SC RI
Ladder-tailed Nightjar	*Hydropsalis climacocerca*	-	U	-	U	RI
Swifts	**Apodidae (4)**					
White-collared Swift	*Streptoprocne zonaris*	'01	U	C	U	MF LF SC
Short-tailed Swift	*Chaetura brachyura*	-	F	U	F	LF SC HU
Band-rumped Swift	*Chaetura spinicauda*	'98 '01	C	F	C	LF RI
Gray-rumped Swift	*Chaetura cinereiventris*	'98 '01	-	-	-	LF MF RI
Hummingbirds	**Trochilidae (7)**					
Eastern Long-tailed Hermit	*Phaethornis superciliosus*	'98 '01	U	U	U	LF MF
Straight-billed Hermit	*Phaethornis bourcieri*	'98	U	U	U	LF MF
Reddish Hermit	*Phaethornis ruber*	'98 '01	U	U	U	LF
Gray-breasted Sabrewing	*Campylopterus largipennis*	'98 '01	S	-	-	LF
White-necked Jacobin	*Florisuga mellivora*	'98 '01	-	-	-	LF RI
Fork-tailed Woodnymph	*Thalurania furcata*	'98	U	U	U	LF MF
Black-eared Fairy	*Heliothryx aurita*	'98	-	-	-	LF
Trogons	**Trogonidae (4)**					
White-tailed Trogon	*Trogon viridis*	'98 '01	F	U	F	LF
Violaceous Trogon	*Trogon violaceus*	'98	U	U	U	LF
Black-throated Trogon	*Trogon rufus*	'98	U	U	U	LF
Black-tailed Trogon	*Trogon melanurus*	'98	U	U	U	LF

continued

Common name	Scientific name	Finch's surveys	Pobawau Creek	Cacique Mountain Forest	Cacique Lowland Forest	Habitats
Kingfishers	**Alcedinidae (5)**					
Ringed Kingfisher	*Ceryle torquata*	'98 '01	F	-	F	RI FW MN
Amazon Kingfisher	*Chloroceryle amazona*	'98 '01	F	-	F	RI FW MN
Green Kingfisher	*Chloroceryle americana*	'98 '01	F	-	F	RI FW MN
Green-and-rufous Kingfisher	*Chloroceryle inda*	'98	S	-	S	RI FW LF
American Pygmy Kingfisher	*Chloroceryle aenea*	-	S	-	S	RI LF FW
Motmots	**Momotidae (1)**					
Blue-crowned Motmot	*Momotus momota*	'98 '01	U	-	U	LF
Jacamars	**Galbulidae (5)**					
Brown Jacamar	*Brachygalba lugubris*	'98 '01	F	-	U	RI SC
Yellow-billed Jacamar**	*Galbula albirostris*	-	S	S	S	LF
Paradise Jacamar	*Galbula dea*	-	U	-	U	LF RI
Green-tailed Jacamar	*Galbula galbula*	'98 '01	-	-	-	LF SC
Great Jacamar	*Jacamerops aurea*	'98	-	-	-	LF
Puffbirds	**Bucconidae (6)**					
Spotted Puffbird	*Bucco tamatia*	-	U	-	U	SC LF
Collared Puffbird	*Bucco capensis*	-	U	-	U	LF
White-necked Puffbird	*Notharchus macrorhynchos*	'01	-	-	-	LF
Pied Puffbird	*Notharchus tectus*	'01	-	-	-	LF
Black Nunbird**	*Monasa atra*	'98 '01	S	-	F	LF
Swallow-wing	*Chelidoptera tenebrosa*	'98 '01	C	-	C	RI SC LF
Barbets, Toucans	**Ramphastidae (6)**					
Black-spotted Barbet	*Capito niger*	'98	-	-	-	LF MF
Green Aracari*	*Pteroglossus viridis*	'01	F	U	F	LF
Black-necked Aracari	*Pteroglossus aracari*	'98 '01	U	-	U	LF
Guianan Toucanet*	*Selenidera culik*	-	U	-	U	LF
Channel-billed Toucan	*Ramphastos vitellinus*	'98 '01	S	-	S	LF
Red-billed Toucan	*Ramphastos tucanus*	'98 '01	U	U	U	LF
Woodpeckers	**Picidae (10)**					
Golden-spangled Piculet	*Picumnus exilis*	-	-	-	U	LF
Lineated Woodpecker	*Dryocopus lineatus*	'98 '01	U	-	-	LF SC HU
Yellow-tufted Woodpecker	*Melanerpes cruentatus*	-	U	-	U	LF HU
Golden-collared Woodpecker*	*Veniliornis cassini*	-	O	-	-	LF
Yellow-throated Woodpecker	*Piculus flavigula*	'98	-	-	-	LF
Waved Woodpecker	*Celeus undatus*	-	U	-	U	LF
Cream-colored Woodpecker	*Celeus flavus*	'98 '01	-	-	O	LF
Ringed Woodpecker	*Celeus torquatus*	'98 '01	-	-	O	LF
Red-necked Woodpecker	*Campephilus rubricollis*	'98	U	U	U	LF MF
Crimson-crested Woodpecker	*Campephilus melanoleucos*	'98 '01	-	-	U	LF HU
Ovenbirds	**Furnariidae (6)**					
McConnell's Spinetail	*Synallaxis macconnelli*	'98 '01	-	-	-	LF MF
Plain-crowned Spinetail	*Synallaxis gujanensis*	'98 '01	-	-	-	SC HU

continued

Common name	Scientific name	Finch's surveys	Pobawau Creek	Cacique Mountain Forest	Cacique Lowland Forest	Habitats
Buff-throated Foliage-gleaner	*Automolus ochrolaemus*	'01	-	-	-	LF
Olive-backed Foliage-gleaner	*Automolus infuscatus*	'98 '01	-	-	-	LF
Chestnut-crowned Foliage-gleaner	*Automolus rufipileatus*	'01	-	-	-	RI LF
Plain Xenops	*Xenops minutus*	'98	U	-	-	LF
Woodcreepers	**Dendrocolaptidae (10)**					
Plain-brown Woodcreeper	*Dendrocincla fuliginosa*	'01	U	-	U	LF
Wedge-billed Woodcreeper	*Glyphorynchus spirurus*	'98 '01	F	-	F	LF MF
Barred Woodcreeper	*Dendrocolaptes certhia*	'98	-	-	-	LF
Black-banded Woodcreeper	*Dendrocolaptes picumnus*	'98	U	-	U	LF MF
Straight-billed Woodcreeper	*Xiphorhynchus picus*	'98	-	-	-	MN RI SC
Striped Woodcreeper	*Xiphorhynchus obsoletus*	'98 '01	U	-	-	LF RI
Chestnut-rumped Woodcreeper	*Xiphorhynchus pardalotus*	'98	-	-	-	LF MF
Buff-throated Woodcreeper	*Xiphorhynchus guttatus*	'98 '01	U	-	U	LF
Lineated Woodcreeper	*Lepidocolaptes albolineatus*	'98	-	-	-	LF
Curve-billed Scythebill	*Campylorhamphus procurvoides*	-	O	-	-	LF
Typical Antbirds	**Thamnophilidae (28)**					
Fasciated Antshrike	*Cymbilaimus lineatus*	'98 '01	-	-	-	LF
Great Antshrike	*Taraba major*	'01	U	-	U	HU SC
Black-crested Antshrike	*Sakesphorus canadensis*	'98 '01	-	-	-	SC MN
Mouse-colored Antshrike	*Thamnophilus murinus*	'98 '01	U	U	U	LF
Amazonian Antshrike	*Thamnophilus amazonicus*	'98 '01	U	-	U	LF RI
Spot-winged Antshrike	*Pygiptila stellaris*	'98	-	-	-	LF
Dusky-throated Antshrike	*Thamnomanes ardesiacus*	'98	-	-	-	LF
Cinereous Antshrike	*Thamnomanes caesius*	'98	O	-	-	LF
Pygmy Antwren	*Myrmotherula brachyura*	'98 '01	U	-	U	LF RI
Streaked Antwren	*Myrmotherula surinamensis*	'98 '01	-	-	-	RI LF
Rufous-bellied Antwren*	*Myrmotherula guttata*	'98	-	-	-	LF
White-flanked Antwren	*Myrmotherula axillaris*	'98 '01	U	-	U	LF SC
Long-winged Antwren	*Myrmotherula longipennis*	'98	-	-	-	LF
Gray Antwren	*Myrmotherula menetriesii*	'98 '01	U	-	U	LF
Spot-tailed Antwren**	*Herpsilochmus sticturus*	'98	-	-	-	LF
Todd's Antwren*	*Herpsilochmus stictocephalus*	'98 '01	-	-	-	LF
Dot-winged Antwren	*Microrhopias quixensis*	'98 '01	-	-	-	LF
Gray Antbird	*Cercomacra cinerascens*	'98 '01	-	-	-	LF
Dusky Antbird	*Cercomacra tyrannina*	'98	U	-	F	LF SC HU
Blackish Antbird	*Cercomacra nigrescens*	'98 '01	U	-	U	RI LF
White-browed Antbird	*Myrmoborus leucophrys*	'98 '01	-	-	-	LF RI HU
Warbling Antbird	*Hypocnemis cantator*	'98	F	-	F	LF
Black-chinned Antbird	*Hypocnemoides melanopogon*	'98 '01	U	-	U	LF RI
Black-headed Antbird**	*Percnostola rufifrons*	-	U	-	U	LF
Spot-winged Antbird	*Percnostola leucostigma*	'98	U	-	U	LF MF
Scale-backed Antbird	*Hylophylax poecilinota*	-	U	-	U	LF

continued

Common name	Scientific name	Finch's surveys	Pobawau Creek	Cacique Mountain Forest	Cacique Lowland Forest	Habitats
Wing-banded Antbird	*Myrmornis torquata*	-	U	-	-	LF
White-plumed Antbird	*Pithys albifrons*	-	O	-	-	LF
Ground Antbirds	**Formicariidae (4)**					
Rufous-capped Antthrush	*Formicarius colma*	'98 '01	-	-	-	LF
Black-faced Antthrush	*Formicarius analis*	'98 '01	-	-	U	LF
Spotted Antpitta	*Hylopezus macularius*	'98	-	-	U	LF
Thrush-like Antpitta	*Myrmothera campanisona*	'98 '01	-	-	U	LF
Tyrant Flycatchers	**Tyrannidae (34)**					
White-lored Tyrannulet	*Ornithion inerme*	'98 '01	-	-	-	LF RI SC
Southern Beardless-Tyrannulet	*Camptostoma obsoletum*	'98 '01	-	-	-	SC LF
Yellow Tyrannulet	*Capsiempis flaveola*	'98	-	-	-	LF SC
Yellow-crowned Tyrannulet	*Tyrannulus elatus*	'98 '01	F	-	F	LF SC
Elaenia	*Elaenia* sp.	-	-	-	O	-
Forest Elaenia	*Myiopagis gaimardii*	'98 '01	-	-	-	LF
Yellow-crowned Tyrannulet	*Myiopagis flavivertex*	'98 '01	-	-	-	RI LF
McConnell's Flycatcher	*Mionectes macconnelli*	'98	-	-	-	LF MF
Slender-footed Tyrannulet	*Zimmerius gracilipes*	'98	U	-	U	LF
Short-tailed Pygmy-Tyrant	*Myiornis ecaudatus*	'98 '01	-	-	-	LF
Double-banded Pygmy-Tyrant	*Lophotriccus vitiosus*	'98 '01	S	-	-	LF
Common Tody-Flycatcher	*Todirostrum cinereum*	'98	U	-	U	SC HU
Painted Tody-Flycatcher*	*Todirostrum pictum*	'01	-	-	-	LF
Ringed Antpipit	*Corythopis torquata*	'98	-	-	-	LF
Rufous-tailed Flatbill	*Ramphotrigon ruficauda*	'98 '01	-	-	-	LF
Yellow-margined Flycatcher	*Tolmomyias assimilis*	'98 '01	-	-	-	LF
Gray-crowned Flycatcher	*Tolmomyias poliocephalus*	'98 '01	-	-	-	LF
Royal Flycatcher	*Onychorhynchus coronatus*	-	O	-	-	LF
Ruddy-tailed Flycatcher	*Terenotriccus erythrurus*	-	U	-	U	LF
Sulphur-rumped Flycatcher	*Myiobius barbatus*	'98	-	-	-	LF
Drab Water-Tyrant	*Ochthornis littoralis*	'98 '01	-	-	-	RI
Amazonian Black-Tyrant	*Knipolegus poecilocercus*	-	F	-	F	RI
Cinnamon Attila	*Attila cinnamomeus*	'98 '01	-	-	U	LF SC RI
Bright-rumped Attila	*Attila spadiceus*	'01	-	-	-	LF MF
Sirystes	*Sirystes sibilator*	'98	-	-	-	LF
Grayish Mourner	*Rhytipterna simplex*	'98 '01	-	-	-	LF
Short-crested Flycatcher	*Myiarchus ferox*	'98 '01	-	-	-	LF SC
Rusty-margined Flycatcher	*Myiozetetes cayanensis*	'01	-	-	-	HU SC RI
Yellow-throated Flycatcher**	*Conopias parva*	'98 '01	-	-	U	LF
Tropical Kingbird	*Tyrannus melancholicus*	'98 '01	F	-	U	SC HU SV
Thrush-like Mourner	*Schiffornis turdinus*	'98	-	-	-	LF MF
White-winged Becard	*Pachyramphus polychopterus*	'01	U	-	U	LF SC
Black-capped Becard	*Pachyramphus marginatus*	'98 '01	-	-	-	LF
Black-tailed Tityra	*Tityra cayana*	'98 '01	O	-	-	LF

continued

Common name	Scientific name	Finch's surveys	Pobawau Creek	Cacique Mountain Forest	Cacique Lowland Forest	Habitats
Cotingas	**Cotingidae (4)**					
Screaming Piha	*Lipaugus vociferans*	'98 '01	C	U	C	LF
Guianan Red-Cotinga	*Phoenicircus carnifex*	-	O	-	-	LF
Pompadour Cotinga	*Xipholena punicea*	-	-	-	O	LF
Purple-throated Fruitcrow	*Querula purpurata*	'98	U	U	U	LF
Manakins	**Pipridae (4)**					
White-throated Manakin*	*Corapipo gutturalis*	'98	-	-	-	MF LF
White-crowned Manakin	*Pipra pipra*	'98	U	U	U	LF MF
Golden-headed Manakin	*Pipra erythrocephala*	'98	U	U	U	LF
White-fronted Manakin*	*Lepidothrix serena*	-	-	O	-	LF MF
Vireos	**Vireonidae (6)**					
Red-eyed Vireo	*Vireo olivaceus*	'98 '01	-	-	-	LF
Lemon-chested Greenlet	*Hylophilus thoracicus*	'98 '01	-	-	-	LF
Buff-cheeked Greenlet	*Hylophilus muscicapinus*	'98 '01	-	-	-	LF MF
Tawny-crowned Greenlet	*Hylophilus ochraceiceps*	'98	-	-	-	LF
Slaty-capped Shrike-Vireo	*Vireolanius leucotis*	'98	-	-	-	LF MF
Rufous-browed Peppershrike	*Cyclarhis gujanensis*	'98 '01	-	-	-	SC HU
Swallows	**Hirundinidae (7)**					
Gray-breasted Martin	*Progne chalybea*	'98	F	-	-	HU SC
Brown-chested Martin	*Progne tapera*	'98	F	-	U	SV SC RI
White-winged Swallow	*Tachycineta albiventer*	'98 '01	C	-	C	RI FW
White-banded Swallow	*Atticora fasciata*	'98 '01	C	-	C	RI
Southern Rough-winged Swallow	*Stelgidopteryx ruficollis*	-	-	C	U	FW RI SC
Bank Swallow	*Riparia riparia*	-	U	-	-	SV HU FW
Barn Swallow	*Hirundo rustica*	'01	-	-	-	SV HU SC
Wrens	**Troglodytidae (2)**					
Coraya Wren	*Thryothorus coraya*	'98 '01	F	-	-	LF
Buff-breasted Wren	*Thryothorus leucotis*	'98 '01	-	-	-	LF RI SC
Gnatwrens, Gnatcatchers	**Polioptilidae (2)**					
Long-billed Gnatwren	*Ramphocaenus melanurus*	'98 '01	O	-	-	LF
Tropical Gnatcatcher	*Polioptila plumbea*	'98 '01	-	-	-	SC LF
Thrushes	**Turdidae (2)**					
Cocoa Thrush	*Turdus fumigatus*	'98 '01		-	U	LF
White-necked Thrush	*Turdus albicollis*	'98	-	-	-	LF MF
Wood Warblers	**Parulidae (2)**					
River Warbler	*Basileuterus rivularis*	-	F	-	F	LF RI
Rose-breasted Chat	*Granatellus pelzelni*	'98	-	-	-	LF
Bananaquits	**Coerebidae (1)**					
Bananaquit	*Coereba flaveola*	'98 '01	F	-	U	LF SC HU
Tanagers	**Thraupidae (14)**					
Guira Tanager	*Hemithraupis guira*	'01	-	-	-	LF
Yellow-backed Tanager	*Hemithraupis flavicollis*	'98	-	-	-	LF

continued

Common name	Scientific name	Finch's surveys	Pobawau Creek	Cacique Mountain Forest	Cacique Lowland Forest	Habitats
Hooded Tanager	*Nemosia pileata*	'01	-	-	-	SC
Flame-crested Tanager	*Tachyphonus cristatus*	'98	-	-	-	LF
Blood-red Tanager	*Piranga haemalea*	'01	-	-	-	MF
Silver-beaked Tanager	*Ramphocelus carbo*	'98 '01	F	-	F	SC HU
Blue-gray Tanager	*Thraupis episcopus*	'98 '01	-	-	-	SC HU
Golden-sided Euphonia	*Euphonia cayennensis*	'98 '01	-	-	-	LF RI
Turquoise Tanager	*Tangara mexicana*	'98 '01	-	-	-	LF HU
Spotted Tanager	*Tangara punctata*	'98	-	-	-	LF MF
Bay-headed Tanager	*Tangara gyrola*	'98	O	-	-	MF LF
Blue Dacnis	*Dacnis cayana*	-	F	U	U	LF SC
Purple Honeycreeper	*Cyanerpes caeruleus*	'98	-	-	-	LF MF
Swallow-Tanager	*Tersina viridis*	'98 '01	-	-	O	LF RI
Emberizine Finches	**Emberizidae (2)**					
Red-capped Cardinal	*Paroaria gularis*	'98	F	-	F	RI SC
Pectoral Sparrow	*Arremon taciturnus*	'98	-	-	-	LF
Grosbeaks, Saltators	**Cardinalidae (3)**					
Slate-colored Grosbeak	*Saltator grossus*	'98 '01	U	-	U	LF
Buff-throated Saltator	*Saltator maximus*	'98 '01	-	-	-	SC LF
Yellow-green Grosbeak	*Caryothraustes canadensis*	-	-	U	U	LF
New World Blackbirds	**Icteridae (5)**					
Giant Cowbird	*Scaphidura oryzivora*	'98 '01	-	-	-	LF SC HU
Moriche Oriole	*Icterus chrysocephalus*	-	U	-	U	SC PA HU
Yellow-rumped Cacique	*Cacicus cela*	'98 '01	O	-	U	RI LF HU
Crested Oropendola	*Psarocolius decumanus*	'98 '01	U	U	U	LF
Green Oropendola	*Psarocolius viridis*	-	U	-	U	LF MF

Appendix 5

Non-volant mammals recorded in the Kanuku Mountains, Guyana

Jim Sanderson

Nonvolant mammals observed, identified by track or sound, or photographed during this expedition and presumed to occur.

OPC = observed at Pobawau Creek site
OCM = observed at Cacique Mountain site
PPC = photographed at Pobawau Creek site
PCM = photographed at Cacique Mountain site
PC = personal communication with local guides
1 = Emmons and Feer (1990)
2 = Agriconsulting (1993)
3 = Parker, et al. (1993)
4 = Nowak (1991)
5 = Lim and Norman (2002)

Taxa	Common Name	Evidence
Marsupialis		
Didelphidae		
Didelphis marsupialis	Common opossum	PPC
Didelphis albiventris	White-eared opossum	1,3
Caluromys lanatus	Western woolly opossum	3
Caluromys philander	Bare-tailed woolly opossum	1
Philander opossum	Common gray four-eyed opossum	OPC, PPC, PCM
Chironectes minimus	Water opossum	1
Metachirus nudicaudatus	Brown four-eyed opossum	1,3
Micoureus demerarae	Woolly mouse opossum	1,3
Marmosops parvidens	Delicate slender mouse opossum	1,3
Marmosa murina	Murine mouse opossum	3
Monodelphis brevicaudata	Red-legged short-tailed opossum	1,3
Edentata		
Bradypodidae		
Bradypus tridactylus	Pale throated three-toed sloth	PC
Choloepidae		
Choloepus didactylus	Southern two-toed sloth	PC

continued

Taxa	Common Name	Evidence
Myrmecophagidae		
Tamandua tetradactyla	Collared tamandua	PC
Cyclopes didactylus	Pygmy anteater	PC
Myrmecophaga tridactyla	Giant anteater	PC
Dasypodidae		
Priodontes maximus	Giant armadillo	PPC
Dasypus novemcinctus	Nine-banded long-nosed armadillo	PC
Dasypus kappleri	Great long-nosed armadillo	PCM
Cabassous unicinctus	Southern naked-tailed armadillo	1
Primates		
Callitrichidae		
Saguinus midas	Golden-handed or midas tamarin	PC
Cebidae		
Saimiri sciureus	Common squirrel monkey	OPC, OCM
Cebus apella	Brown capuchin monkey	OPC, OCM
Cebus olivaceus	Wedge-capped capuchin	OPC, OCM
Pithecia pithecia	White-faced saki	PC
Chiropotes santanas	Brown bearded saki	OPC, OCM
Alouatta seniculus	Red howler monkey	OPC, OCM
Ateles paniscus	Black spider monkey	OPC, OCM
Carnivora		
Canidae		
Speothos venaticus	Bush dog	PC
Cerdocyon thous	Crab-eating fox	2,3
Urocyon cinereoargenteus	Grey fox	2
Procyonidae		
Procyon cancrivorus	Crab-eating raccoon	PC
Potos flavus	Kinkajou	OCM
Nasua nasua	South American coati	2
Mustelidae		
Galictis vittata	Grison	1
Eira barbara	Tayra	1
Lutra longicaudis	Southern river otter	PC
Pteronura brasiliensis	Giant river otter	PC
Felidae		
Felis concolor	Puma	PC
Felis jaguarundi	Jaguarundi	PC
Leopardus pardalis	Ocelot	PPC
Leopardus wiedi	Margay	PPC
Leopardus tigrina	Oncilla	1-3
Panthera onca	Jaguar	OPC

continued

Taxa	Common Name	Evidence
Artiodactyla		
Cervidae		
Mazama gouazoubira	Gray brocket deer	PPC
Mazama americana	Red brocket deer	PPC, PCM
Odocoileus virginianus	White-tailed deer	2,3
Tayassuidae		
Tayassu pecari	White-lipped peccary	PCM, PPC
Tayassu tajacu	Collared peccary	PCM, PPC
Perissodactyla		
Tapiridae		
Tapirus terrestris	Brazilian tapir	OCM
Rodentia		
Sciuridae		
Sciurus aestuans	Guianan squirrel	1
Erithizontidae		
Coendou prehensilis	Brazilian porcupine	PC
Caviidae		
Cavia guianae	Guinea pig	4
Hydrochaeridae		
Hydrochaeris hydrochaeris	Capybara	OCM
Agoutidae		
Agouti paca	Paca	PPC
Dasyproctidae		
Dasyprocta cristata	Red-rumped agouti	PPC, PCM
Myoprocta exilis	Green acouchi	PPC, PCM
Echimyidae		
Procechimys guyanensis	Spiny rat	PCM
Oryzomys megacephalus	Large-headed rice rat	5